Open RAN Explained

Open RAN Explained

The New Era of Radio Networks

Jyrki T. J. Penttinen
Technical Manager, GSMA North America

Michele Zarri
Management Consultant, UK

Dongwook Kim
Technical Officer, European Telecommunications Standards Institute (ETSI), France

Library of Congress Cataloging-in-Publication Data Applied for
[Hardback ISBN: 9781119847045]

Cover Design and Image: © Wiley

Set in 9.5/12.5pt STIXTwoText by Straive, Chennai, India

Contents

Author Biographies

Jyrki T.J. Penttinen has worked on mobile telecommunications in Finland, Spain, Mexico, and the United States since 1994. His past employers include Telia Sonera, Nokia, G+D Mobile Security Americas, and Syniverse, and he is with GSMA North America Technology Team at present. He is experienced in research and operational activities such as planning, optimization, measurements, system architectures, and services. Dr. Penttinen is also an active lecturer and has authored various books on telecommunication technologies.

Michele Zarri is an independent management consultant. He started his career in Fujitsu R&D working on layer 1 of WCDMA before moving to Deutsche Telekom representing the company in 3GPP. Michele served two terms as 3GPP TSG SA WG1 chairman and has been rapporteur of several specifications and work items. In 2015, Michele joined the GSMA as technical director of the 5G projects.

Dongwook Kim is Specifications Manager of the 3GPP, the secretary of 3GPP CT3 working group, and the work plan coordinator of 3GPP TSG CT. He has worked in the industry for 11 years and his past employers include Korea Telecom (KT), GSM Association (GSMA), and Telecom Infra Project (TIP). His career has focused on promoting latest telecom technologies and serving as the industry think tank.

Preface

The primary intention of the standardization of the mobile communications systems is to ensure as good interoperability between the system components as practically feasible. This principle has provided the mobile network operators with means to design and deploy their networks relying upon different equipment vendor solutions for the radio and core segments. Nevertheless, in practice, the network components within the radio access network (RAN) have been typically so tightly integrated by each vendor that it has been all but impossible to mix and match different radio network equipment providers' solutions within the same RAN.

Industry has thus decided to put further efforts, building upon the 3GPP specifications, to extend the current architectures to cover more standardized interfaces also within the RAN itself. This allows operators to disaggregate the RAN into a set of interoperable components. Such disaggregation has in turn facilitated the emergence of totally new stakeholders that are no longer required to be able to provide the full RAN stack but can instead focus on some of its components. The abstraction of these contact points makes the new RAN environment more transparent and interoperable and can have a positive impact on the business of all the involved parties in terms of increased number of available solutions as well as potentially bring more innovation. Operators are provided with more options to choose from to evolve their radio access segment as well as deploy bespoke solutions for some specific scenarios.

The new environment is still in relatively initial phase regardless of the very active efforts the telco industry has invested to evolve the concept, but the Open RAN is getting increasingly real now. There are already several examples of practical deployments, while the standardization efforts continue detailing adequate solutions. The effort is not, however, completely straightforward, and some challenges will require more time to be successfully addressed. For example, new RAN component concepts such as Radio Intelligent Controller (RIC) is not expected to reach its full potential initially, while operators learn how to leverage the underlying machine learning models and various use cases that artificial intelligence solutions can bring. Moreover, the move toward virtualization of the RAN will require operators to become familiar with orchestration strategies.

Evolved measurement techniques, testing, and processes are needed to ensure the new concept works adequately prior to production and deployment while ensuring adequate performance through the rest of the lifetime of the networks. It is important to note that RAN often accounts for around 70% of the CAPEX of a typical operator.

Yet another challenge, regardless of the increasing number of references becoming public, has been the lack of concrete publications detailing the concept, its more concrete possibilities and challenges, and the ways to deploy the Open RAN in practice.

This book answers to the need by presenting the Open RAN concept based on the latest specifications and information sources and walks the readers through some of the very key aspects that the ecosystem needs to understand in the functioning, deployment, and operation of the Open RAN-based networks.

This effort to summarize sufficiently and concretely the essential between single covers has been challenging due to such fast pace of the development and the lack of adequate references. Our author team is extremely happy to share the result in a form of this book which we hope to serve the ecosystem in our efforts to make sense out of the complex and oftentimes rather fragmented public information sources.

This book is thus a result of rather long exploration of the environment and root sources such as key specifications of the 3GPP and Open RAN Alliance. We hope this effort benefits the mobile communications ecosystem to learn more about the Open RAN, the topic that has rather realistic prospects to become a highly significant – perhaps even elemental – part of the modern telecom systems, and that is expected to work as important driver for generating new business through evolving ecosystem and new stakeholders.

"Mobile telecommunication systems have been an integral part of people's lives for such a long time that only few of us would really like to return to the era of sole fixed telephony. Having seen the development of the wireless industry from many points of view since 1980s through technical engineering career, starting off with radio network measurements of the very first generation, and working posteriorly with operators, manufacturers, security and roaming providers, and membership organizations, I have been fortunate to witness some of the key breakthrough moments of the wireless industry. Some examples include the commercialization of the 2G in Finland back in 1991, the standardization of the first truly IP-based mobile data service, General Packet Radio Service (GPRS), the pre-commercial field testing of the 3rd Generation UMTS (Universal Mobile Telecommunications System), and the takeover of the 4G LTE that currently represents the dominating radio technology. This journey is becoming increasingly interesting as the 5G, which I started to research from the specifications prior to its commercial readiness, is maturing firmly and starts offering advanced features and functions such as network slicing and other 5G SA capabilities also in practice.

Based on these personal experiences, I realize the development of mobile communication systems is a constant effort that materializes in cycles of each decade as completely new generation becomes commercially available. Each new generation tackle important lessons learned that the ecosystem has gained through the previous ones. I also reckon that – apart from the actual deployment and commercial start of the new generation – it is hard to think of much more significant and groundbreaking moments than the gradual availability of the Open RAN concept. This new concept has also provided a fantastic opportunity to learn and share latest knowledge, including the security and testing specifications of the Open RAN Alliance."

Jyrki T. J. Penttinen, Atlanta, Georgia, USA, 2023

"I have been involved in mobile standardization for more than 20 years and I am sure that the rise of the Open RAN "movement" will be remembered as a major milestone along with the creation of 3GPP, the selection of WCDMA as radio technology for 3G, the battle between LTE and WiMAX during the design of 4G and the introduction of service based architecture in 5G.

Besides addressing well-known shortcomings of the existing RAN architecture, Open RAN drive to transfer to tangible benefits of virtualization, separation of hardware and software and disaggregation that have proven their worth in the IT world, creates the premises for establishing a healthier supply chain, foster innovation and ultimately make the RAN more affordable. A cheaper, better RAN will bring societal benefits such as reducing the digital divide as well as economic benefits by unlocking new commercial opportunity.

While the jury is still out as to whether all these promises will materialize and challenges will be overcome, it is clear that the efforts of TIP and O-RAN Alliance have not gone unnoticed and acted as a wake-up call for the established vendors who might have been too slow in adopting new technologies and paradigms.

Moving forward, the most desirable outcome from my point of view is that the principles, components and specifications developed in Open RAN converge in 3GPP avoiding a divergence of mobile communication system standards that may damage in the long run the economies of scale and pace of innovation. Precedents exist of ideas generated outside the "mainstream" that were contributed to and implemented in 3GPP: the IP Multimedia Subsystem initially devised by 3G. IP and the RAN split first introduced by X-RAN being notable examples. Such convergence of Open RAN and 3GPP would also create the best premises for the development of a successful, global, open 6G."

Michele Zarri, London, UK, 2023

"Modular architecture and separation of integrated layers is already prevalent in our lives. From Lego in toys to our personal computers, we often mix and match components to build what we want when we want it and how we want it to be. It is no surprise that I witnessed, in the start of my career, similar work in the core network side starting with Network Function Virtualization (NFV), leading to a great success that is still on-going within the Industry Specification Group NFV at ETSI. This shows that the principles and the trend of Open RAN is not as complicated and strange as it initially seems, it is the quest of the network operator (whether it be traditional network operators or emerging alternative network models in the 5G era) to flexibly and optimally deploy and operate its network.

However, openness can be tricky in that the user/customer needs a degree of knowledge to fully exploit the potential. Taking the Lego example, an average child playing with the toy will not be able to build a gigantic and magnificent masterpiece that you would find in Lego Land, let alone the Lego toy series that manias display in their glass cupboards. In this respect, I believe that this book will set a stepping stone for you to be able to play with Open RAN like the Lego manias do with their Lego toys. Of course, you should not limit yourself to Open RAN per se as mobile networks is a much more complex topic and should be aware of the relevant 3GPP work that Open RAN is based on."

Dongwook Kim, Sophia Antipolis, France, 2023

Acknowledgments

This book is a result of countless hours of exploration of mobile communications resources through specifications and other available information sources, discussions with our peers, as well as ideation and manuscript drafting, that all were essential steps for us to be able to write down the contents of this book. It has been a challenging task, yet highly rewarding as we wrote this book to provide ecosystem with concrete ways to learn more on the subject.

Our author team would like to acknowledge all the ones we had possibility to discuss the topic throughout this effort, including our colleagues and peers at GSMA and 3GPP. We also appreciate all the support and patience of our close families during this work.

Finally, this book would not be reality without the firm, yet gentle guidance and coordination of the Wiley team. Thank you so much Nandhini Karuppiah, Sandra Grayson, Becky Cowan, and all the ones within Wiley involved in this effort directly and in supporting roles. It has been a pleasure to work under such a professional guidance and friendly spirit.

Abbreviations

1G	first generation of mobile communications
2G	second generation of mobile communications
3G	third generation of mobile communications
3GPP	third generation partnership project
4G	fourth generation of mobile communications
5G	fifth generation of mobile communications
5GS	5G system
6G	sixth generation of mobile communications
A/V	audio/video
AAL	accelerator abstraction layer
AAL	ATM adaptation layer
ACPI	advanced configuration and power interface
ADC	analogue to digital conversion
AI	artificial intelligence
AICPA	American Institute of Certified Public Accountants
ALD	antenna line device
AMF	access and mobility management function (5G)
ANR	automatic neighbor relation
AP	application plane
API	application programming interface
AR	augmented reality
ARIB	Association of Radio Industries and Businesses, Japan
ASIC	application-specific integrated circuit
ATIS	Alliance for Telecommunications Industry Solutions (North America)
BBU	baseband unit
BoM	bill of material
BSS	business support system
BVLOS	beyond visual LOS
CA	carrier aggregation
CA	certificate authority
CAGR	cumulative annual growth rate
CAPEX	capital expenditure
CCSA	China Communications Standards Association

CCTV	closed circuit television
CD/CT	continuous deployment/continuous testing
CDMA	code division multiple access
CI	cloud infrastructure
CI/CD	Continuous integration (CI) and continuous deployment (CD)
CICA	Canadian Institute of Chartered Accountants
CII	core infrastructure initiative (LF)
CM	configuration management
CMP	certificate management protocol (PKI)
CN	core network
CNF	cloud network function
CoMP	coordinated multi-point (transmission/reception)
COTS	commercial off-the-shelf hardware
CP	circular polarization
CP	control plane
CP	cyclic prefix
CPRI	common public radio interface
CPU	central processing unit
C-RAN	cloud RAN
CRC	cyclic redundancy check
C-SCRM	cybersecurity supply chain risk management
CSP	cloud service provider
CSRIC	Communications Security, Reliability, and Interoperability Council
CT	core network and terminals (3GPP TSG)
CU	centralized unit
CU-CP	CU split to control plane
CUPS	control and user plane separation
CU-UP	CU split to user plane
DIFI	digital intermediate frequency interoperability
DL	downlink
DMS	deployment management services
DoC	Department of Commerce (USA)
DP	data plane
DRB	data radio bearer
DSRC	dedicated short-range communication
DSS	dynamic spectrum sharing
DTLS	datagram transport layer security
DU	distributed unit
DUT	device under testing
E2E	end-to-end
EM	element management
EMS	element management system
eNB	Evolved NodeB (4G)
ENISA	European Union Agency for Cybersecurity
eNodeB	LTE node (4G), *see* eNB

EPC	evolved packet core (4G)
eRE	evolved radio equipment
eREC	evolved radio equipment controller
ERP	enterprise resource planning
eSIM	embedded SIM
EST	enrollment over secure transport protocol
ETSI	European Telecommunications Standards Institute
EU	European Union
E-UTRAN	evolved universal terrestrial radio access network (4G)
FCAPS	fault, configuration, accounting, performance, and security management
FCC	Federal Communications Commission (USA)
FDMA	frequency division multiple access
FFT	fast Fourier transformation
FH	Fronthaul
FM	fault management
FOCOM	federated O-cloud orchestration and management
FPGA	field programmable gate array
FR	frequency range
FRMCS	future railway mobile communication system
GEO	geostationary orbit
gNB	next-generation NodeB (5G)
GoB	grid of beams (MIMO)
GPP	general-purpose processor
GPS	global positioning system
GPU	graphics processing unit
GSM	global system for mobile communications (2G)
GSMA	GSM association
GSM-R	GSM-Railway
GTP	GPRS transfer protocol
GTP-U	GPRS transfer protocol user plane
GUI	graphical user interface
HARQ	hybrid automatic repeat request
HD	high definition
HDLC	high-level data link control
HLS	high-level split
HO	handover
HSM	hardware security module
HW	hardware
i/f	interface
IaaS	infrastructure as a service
ICT	Information and Communication Technology
IDS	intrusion detection system
IEEE	Institute of Electrical and Electronics Engineers
IEFG	Industry Engagement Focus Group (O-RAN Alliance)
IF	intermediate frequency

iFFT	inverted fast Fourier transform
IMF	infrastructure management framework
IMS	infrastructure management services
IMS	IP multimedia subsystem
IoT	internet of things
IOT	interoperability testing
IP	internet protocol
IPS	intrusion prevention system
IPSec	internet protocol security
IQ	in-phase (I) and quadrature (Q) component of sinusoids of amplitude modulation
ISG	Industry Specification Group (ETSI)
ISL	inter-satellite link
ISO	International Organization for Standardization
IT	information technology
ITU	International Telecommunications Union
ITU-T	ITU Telecommunication Standardization Sector
KPI	key performance indicator
L1	layer 1
LCM	life cycle management
LEO	low earth orbit
LF	linux foundation
LHCP	left-hand circular polarization
LLS	low-level split
LMF	location management function
LOS	line-of-sight
LP	linear polarization
LSB	least significant bit
LTE	long-term evolution (4G radio network)
LTE-A	LTE advanced (4G)
MAC	medium access control
MANO	NFV management and orchestration (ETSI)
MC	mission critical
MCG	Master Cell Group (4G, 5G)
MCPTT	mission critical push-to-talk
MEC	multi-access edge computing
MEO	medium earth orbit
MIMO	multiple-input multiple-output
mIoT	massive IoT
ML	machine Learning
MME	mobility management entity (4G)
mMIMO	massive MIMO
MNO	mobile network operator
MOCN	multi operator core network
MORAN	multi operator radio access network

MoU	memorandum of understanding
MP	management plane
MRO	mobility robustness optimization
MSB	most significant bit
MU-MIMO	multi-user MIMO
NaaS	network-as-a-service
NB	NodeB (3G)
NB-IoT	narrow-band IoT
NE	network element
NE-DC	NR/E-UTRA dual carrier (4G, 5G)
NESAS	network equipment security assurance scheme (GSMA)
NETCONF	network configuration protocol
NF	network function
NFO	network function orchestrator
NFV	network function virtualization
NFVI	NFV infrastructure
ng-eNB	evolved 4G nodeB supporting split model
NGMN	next generation mobile network
NG-RAN	next generation radio access network
nGRG	Next Generation Focus Group (O-RAN alliance)
NIB	network information database
NIST	National Institute of Standards and Technology (USA)
NLOS	non-line of sight
NMS	network management system
NPN	non-public network
NR	new radio (5G)
nRT-RIC	near-real-time RIC
NSA	non-standalone (4G, 5G)
NSI	network slice instance
NSSI	network slice subnet instance
NTIA	National Telecommunications and Information Administration (USA)
NTN	non-terrestrial network
OAM	operations and maintenance
OAuth	open authentication
OCG	Organized Crime Group
O-Cloud	open cloud (O-RAN alliance)
O-CU	Centralized Unit defined by O-RAN Alliance
O-CU-CP	O-CU control plane
O-CU-UP	O-CU user plane
O-DU	distributed unit defined by O-RAN alliance
O-eNB	O-RAN evolved radio access node (4G)
OFDM	orthogonal frequency division multiplex
OFH	open fronthaul
OMG	OpenRAN MoU Group
ONAP	open networking automation platform

ONF	The Open Networking Foundation
ONOS	open network operating system
OP	organizational partner
OPEX	operating expenses
O-RAN	open RAN alliance
O-RU	radio unit defined by O-RAN alliance
OS	operating system
OSFG	Open Source Focus Group (O-RAN alliance)
OSI	open system interconnection
OSM	open-source MANO
OSS	open-source software
OSS	operations support system (RAN)
OTIC	open testing and integration centre
OWASP	open web application security project
PaaS	platform as a service
PBX	private branch exchange
PDCP	packet data convergence protocol
PG	Project Group (Telecom Infra Project)
PGW	packet gateway
PGW-C	PGW control plane
PHY	physical layer
PKI	public key infrastructure
PLMN	public land mobile network
PM	performance monitoring
PNI-NPN	public network integrated NPN
PoC	proof of concept
PRB	physical resource block
PSTN	public switched telephone network
PTP	point to point
PTT	push-to-talk
QoE	quality of experience
QoS	quality of service
RAN	radio access network
RAN	radio access network (3GPP TSG)
rApp	function managed by the non-real-time RIC
RAT	radio access technology
RE	radio equipment
RE	resource element
REC	radio equipment controller
RF	radio frequency
RHCP	right hand circular polarization
RIA	RAN intelligence automation
RIC	RAN intelligent controllers
RLC	radio link control
R-NIB	RAN network information database

RoE	radio over ethernet (FH protocol)
ROMA	RAN orchestration and management
RRC	radio resource control
RRH	remote radio head
RRM	radio resource management
RRU	remote radio unit
RSP	remote SIM provisioning
RTOS	real-time operating system
RU	radio unit
SA	service and system aspects (3GPP TSG)
SA	standalone (5G)
SaaS	software as a service
SAP	system applications and products
SATCOM	satellite communications
SBOM	software bill of material
SCG	Secondary Cell Group (4G, 5G)
SCTP	stream transmission control protocol
SDAP	service data adaptation protocol
SDFG	Standard Development Focus Group (O-RAN alliance)
SDLC	software development lifecycle
SDN	software-defined networking
SDO	standard developing organization
SD-RAN	software-defined radio access network
Sec	security
SFG	Security Focus Group (O-RAN alliance)
SGW	serving gateway
SGW-C	SGW control plane
SIEM	security information and event management
SIM	subscription identity module
SLA	service level assurance/agreement
SMF	session management function (5G)
SMO	service management and orchestration framework
SMS	short message service
SNCF	Sociedad Nacional de Ferrocarriles Franceses
SNMP	simple network management protocol
SNPN	stand-alone non-public network
SON	self-organizing network
SP	service provider
SRB	signaling radio bearer
SRI	satellite radio interface
SSDF	secure software development framework
SSH	secure shell protocol
SuFG	Sustainability Focus Group (O-RAN alliance)
SW	software
TCO	total cost of ownership

TCP	transmission control protocol
TDD	time division duplex
TDF	traffic detection function
TDF-C	TDF control plane
TDM	time division multiplexing
TDMA	time division multiple access
TIFG	Testing and Integration Focus Group (O-RAN alliance)
TIP	telecom infra project
TLS	transport layer security
TSC	Technical Steering Committee (O-RAN Forum)
TSDSI	Telecommunications Standards Development Society, India
TSG	Technical Specification Group
TTA	Telecommunications Technology Association, South Korea
TTC	Telecommunication Technology Committee, Japan
UAV	uncrewed aerial vehicle
UDP	user datagram protocol
UE	user equipment
UE-NIB	UE network information base
UI	user interface
UIC	International Union of Railways
UL	uplink
UMTS	universal mobile telecommunications system (3G)
UP	user plane
UPF	user plane function (5G)
URLLC	ultra-reliable low latency communications
V2C	vehicle to cloud
V2D	vehicle to device
V2G	vehicle to grid
V2I	vehicle to infrastructure
V2N	vehicle to network
V2P	vehicle to pedestrian
V2V	vehicle to vehicle
V2X	vehicle-to-everything
VLAN	virtual local area network
VM	virtual machine
VNF	virtualized network function
Vo5GS	voice over 5G system
vO-CU	virtual open RAN centralized unit
vO-DU	virtual open RAN distributed unit
VoNR	voice over new radio
VR	virtual reality
vRAN	virtualized RAN
WAP	wireless application protocol
WG	workgroup (O-RAN alliance)
WG	working group (3GPP)

X2AP	X2 application protocol
xApp	near real-time RIC multi-vendor programmable application
XR	extended reality
xRAN	extensible RAN (forum)
ZTA	zero trust architecture

1

Introduction

1.1 Overview

Mobile networks have provided consumers location-independent communications since the 1980s. New mobile communication networks and generations build upon the technical insights and experiences of the earlier versions to provide better performance, capacity, and quality to the end users. The fifth generation is no exception to this evolutionary path.

Although the network architectures have evolved, the basic principles of the radio and core networks have remained largely the same. Traditionally, mobile operators manage a set of network elements and their hardware and software dedicated to their respective functions. Only recently, and largely thanks to the service-based architecture model introduced along with the new fifth generation of mobile communications (5G), the virtualization of network functions has become possible natively in the design. At the same time, the established and new standardization bodies update the traditional models to reflect this evolution. Therefore, the operators have additional options to design and deploy modern networks such as virtual and Open radio access network (Open RAN) models [1].

The radio network is the most critical and challenging part of the communications path of the wireless systems. Some of the reasons for this are the unpredictable characteristics of the radio interface with hard-to-predict variables such as deviations in peak and average load. The radio link budget provides good approximation for the operators to predict the average situation, but the unknown aspects remain, altering the final radio quality. The new concepts enhance the performance of the radio networks as they can adapt more optimally to the varying situations in terms of the offered capacity and achievable performance.

Not only the unpredictable radio interface can be a potential issue, but the radio network architectures and their interfaces represent typically rather closed environment for network infrastructure providers. It has been thus challenging, if not impossible, for new players to join the ecosystem, providing alternative or parallel radio network functions [1].

After several years of conceptual discussions, the Open RAN is now becoming a reality, especially along with the deployment of the second phase of the 5G system. O-RAN Alliance has already produced concrete specifications that complement the ones of the 3GPP to further abstract the radio access network (RAN) and interfaces. This gives room for advanced deployment scenarios involving increasing number of incumbent and nontraditional stakeholders alike that can provide and use truly interoperable modules for the radio networks.

Open RAN Explained: The New Era of Radio Networks, First Edition.
Jyrki T. J. Penttinen, Michele Zarri, and Dongwook Kim.
© 2024 John Wiley & Sons Ltd. Published 2024 by John Wiley & Sons Ltd.

1.2   Readiness of the Ecosystem

1.2.1   Virtualization

The radio networks of the mobile communication systems have enhanced dramatically since the 1G networks became a reality. This evolution path that the 3GPP standardization has driven includes the optimization of the architectural models (the LTE and 5G base stations integrate the control functions that separate elements managed in 2G and 3G) and split models for the control and user plane separation (distributed unit [DU], central unit [CU], and radio unit [RU], respectively).

The benefits of these technologies include optimization of the capacity as operators can rededicate it automatically among their base stations, and they provide advanced ways to expand the networks, thanks to the possibility to dedicate more capacity considering more granular modules instead of a mere base station unit. The 3GPP has also specified virtualized RAN (vRAN) model that provides, in theory, more open environment for the radio network deployment. Nevertheless, the vRAN model still has practical interoperability challenges in offering operators with means to implement common interfaces and functions.

The next step in this evolution is the O-RAN that the O-RAN Alliance has specified. It expands the virtual RAN model of the 3GPP, extends interfaces and technical solutions providing better interoperability between stakeholders, and provides technical means for additional providers of different modules to join the infrastructure.

The Open RAN is already a reality. Yet, the ecosystem lacks information sources to understand the concept and its potential. This book is designed to help operators, equipment manufacturers, service providers, regulators, and educational institutes alike understand and explore the principles and details of the Open RAN.

The recent public news indicates that the interest in Open RAN continues to grow and that the industry works toward networks that can be deployed and operated based on mixing and matching components from different suppliers. As an example, Deutsche Telekom, Orange, Telefónica, and Vodafone Group have signed a Memorandum of Understanding (MoU) to implement Open RAN across Europe.

The move aims to promote the timely deployment of Open RAN technologies and ensure that a strong ecosystem of companies emerges in Europe; Orange has stated that it wants all its network upgrades to be Open RAN by 2025 and puts pressure on traditional equipment suppliers to adapt to new open standards. There are similar examples, e.g. in the United States (advances of Parallel Wireless in Open RAN products and plans of DISH Network to deploy their nationwide 5G network based on the Open RAN paradigm) and Japan (Rakuten is one of the members of the Open RAN Alliance and is considered oftentimes a reference for early-stage multi-vendor RAN deployments).

Not only the standard-setting bodies, device manufacturers, and operators are in a key position to pave the way for the new technology but also the regulators have an important role. As an example, Nokia states in their public source that some countries are looking at Open RAN to introduce new suppliers to the market, while Open RAN can considerably reduce entry barriers for new stakeholders. Furthermore, some countries are looking at Open RAN as an opportunity to drive local 5G innovation.

1.2.2 Industry Forums

Open RAN has had already a long journey. The role of 3GPP has been instrumental in this path, and the resulting technical specifications allow operators to deploy RANs by splitting the RAN structure into the controlled division of the RU, DU, and centralized unit (CU). The 3GPP specifications provide flexibility that optimizes the signaling, capacity, and processing of the RAN in such a way that part of the base station functionalities can be run elsewhere, such as in an edge cloud.

In recent years there have been efforts to extend the 3GPP solutions to enable even more open interfaces within the RAN. This would allow more diverse "mix and match" of RAN components in applying ideally "plug and play" principles. Although there are practical challenges in achieving such a goal, the integration of different vendor solutions within a single RAN area has become a reality.

3GPP has played an essential role in the evolution of mobile systems. As a result, the 3GPP-defined 5G defines a gNB functions split separating the CU and a DU, through the F1 interface. The CU can be split further into a control plane (CU-CP) and user plane (CU-UP), which has resulted in the introduction of the E1 interface. This split model is essential for the further work of the industry to also ensure that the rest of the interfaces are compatible among different vendors.

The currently most important entities, building upon the 3GPP standards for the evolution path for the RAN, are O-RAN Alliance and Telecom Infra Project (TIP). They complement and build definitions upon the split architecture that the 3GPP has facilitated. There are also additional groups that contribute to the development of the O-RAN concept, such as Open Networking Foundation (ONF) and Open RAN Policy Coalition.

The O-RAN Alliance (O-RAN) is committed to evolving RANs with its core principles being intelligence and openness. As stated in [2], it aims to drive the mobile industry toward an ecosystem of innovative, multi-vendor, interoperable, and autonomous RAN, with reduced cost, improved performance, and greater agility. The O-RAN architecture is the foundation for building a vRAN on open hardware and cloud, with embedded AI-powered radio control. The O-RAN provides specifications that complement the 3GPP definitions for the Open RAN concept to ensure the select RAN interfaces are truly interoperable. The O-RAN also provides whitepapers and guidelines through their portal.

The O-RAN specifications present solutions to provide disaggregated, virtualized, and software-based components through open and standardized interfaces. The aim of the O-RAN is to have these interfaces interoperable across multiple vendors building upon the technical principles of 3GPP 4G and 5G RANs, especially on the split models. In practice, the O-RAN focuses on the extension of the 3GPP NR 7-2 split model for the radio network, disaggregating the functions of the base station into a RU, CU, and DU. In addition, the O-RAN specifies the connection to RAN Intelligent Controllers (RIC), covering both near-real-time RIC (nRT-RIC) and non-real-time RIC and including their respective control and policies. These RIC components manage and control the network providing near-real-time performance (10–1000 ms) and non-real-time performance (1 seconds and beyond).

The Telecom Infra Project (TIP) contributes to Open RAN concept. As stated in [3], TIP is a global community of companies and organizations that are driving infrastructure solutions to advance global connectivity. The motivation is to enhance the possibilities for the world's population to access the internet with sufficient quality and to drive consumer and commercial benefits and economic growth. TIP has reasoned that a lack of flexibility in the current solutions and a limited choice in technology providers make it challenging for operators to efficiently build and upgrade networks.

TIP was founded in 2016 to facilitate the cooperation between companies, such as service providers, technology partners, and systems integrators, among other connectivity stakeholders. The work of the project includes the development, testing, and deployment of open, disaggregated, and standards-based solutions. TIP thus adopts the O-RAN Alliance specifications and drives Open RAN initiatives focusing on the enablement of testing and Open RAN deployments by facilitating the discussions among operators, vendors, and system integrators.

More specifically, the Open RAN workgroup of TIP realizes the mission of OpenRAN: that is to accelerate innovation and commercialization in RAN domain with multi-vendor interoperable products and solutions that are easy to integrate into the operator's network and are verified for different deployment scenarios. TIP's OpenRAN program supports the development of disaggregated and interoperable 2G/3G/4G/5G NR RAN solutions based on service provider requirements [4]. The goal of the TIP OpenRAN is to bring the ecosystem together to take a holistic approach toward building next-generation RAN. For this, TIP OpenRAN collaborates with other industry organizations, including GSMA, ONF, and O-RAN Alliance.

The Open Networking Foundation (ONF) is involved in mobile network projects, creating open-source solutions in the rapidly evolving mobile infrastructure space. One of the areas of the ONF is the SD-RAN, which is building open-source components for the mobile RAN space, complementing O-RAN's focus on architecture and interfaces by building and trialing O-RAN-compliant open-source components. The aim of this setup is to foster the creation of multi-vendor RAN solutions and help invigorate innovation across the RAN ecosystem.

The SD-RAN of the ONF develops the Near Real-Time RIC and a set of exemplar xApps for controlling the RAN. This RIC is cloud-native and builds on several of ONF's platforms including the ONOS SDN Controller. The architecture for the SD-RAN near Real-Time RIC will leverage the O-RAN architecture and vision. SD-RAN also follows the O-RAN specification advances, and the SD-RAN project creates new functionality and extensions and feeds the respective learnings back to the O-RAN Alliance with the aim of these extensions to help advance the O-RAN specifications [5].

The Open RAN Policy Coalition is a group of companies formed to promote policies that will advance the adoption of open and interoperable solutions in the RAN to create innovation, spur competition, and expand the supply chain for advanced wireless technologies including 5G [6].

1.2.3 Statistics

The telecom industry is advancing with the practical Open RAN deployments. According to reference [7], iGR Research identified 23 publicly announced mobile network operators

(MNOs) around the world using equipment from multiple vendors who had deployed Open RAN in commercial networks in 2023.

The related business is growing significantly based on the observations of ABI Research that estimates that total spending on O-RAN RUs for the public outdoor macro-cell network will reach US$ 69.5 billion market by 2030 [8]. Furthermore, reference [9] reasons that 5G-based Open RAN deployments will accelerate after 2025, along with various mobile operators stepping onboard, and as the open interface standards evolve and mature sufficiently for the actual interoperability testing. Reference [9] also states that by 2030 there will be 1.3 million O-RAN cells deployed, and their total value is US$ 19.2 billion, as per the Rethink Technology Research's wireless forecasting service – RAN Research report.

As indicated in [10], Parallel Wireless, Mavenir, Altiostar, Samsung, and Radisys are examples of the virtual RAN vendors. The rather large list of present O-RAN RU vendors includes NEC, Fujitsu, Gigatera, Lime Microsystems, Nokia, Airspan, Sercomm, and VANU. Intel, Qualcomm, Xilinx, NVIDIA, and Saankhya Labs are some of the examples of the O-RAN chipset vendors, and the O-RAN system integrator list includes NEC, Rakuten, Everis, Tech Mahindra, and Amdocs.

As for the initial phase of the Open RAN deployment landscape, reference [9] reasons that the early Open RAN adopters have typically been greenfield sites regardless of the region. In the beginning of the deployments, Japan was the forerunner because of the active work of the 5G greenfield operator Rakuten Mobile.

In the United States, DISH Network is an example of early Open RAN adopters along with the greenfield 5G network deployment. The respective early Open RAN adoption is thus free from the potential challenges of the legacy equipment accommodation.

1.2.4 Path Toward Open RAN Networks

The evolution toward the deployment of Open RAN networks can be presented by main phases that are depicted in Figure 1.1. The following list summarizes the key aspects of these phases, as interpreted from references [10] and [11]:

- *Open fronthaul* (stage 1): In this scenario, the baseband unit (BBU) is connected to the RU based on proprietary hardware platforms. This stage provides an open interface between the RU and DU and represents the fronthaul interface covering the RAN split option 7-2x. To truly work in practice, the open fronthaul requires that the participating vendors support it between the RU and DU.
- *Open baseband unit* (stage 2): In this scenario, the RU is connected to DU, which, in turn, connects to CU based on Commercial Off-the-Shelf (COTS) hardware platforms. This scenario is the next step in the evolution toward open interfaces and extends the previous scenario of opening the F1 interface between the DU and CU as per the 3GPP RAN model's higher layer split. In order to provide a true interoperability among CU and DU vendors, they must support the open standard. This scenario helps evolve also the previously applied proprietary hardware approach to allow the use of COTS for the DU and CU.
- *Open, intelligent, and programmable environment* (stage 3): In this scenario, the RU is connected to DU, which connects to RIC, CU-CP, and CU-UP, based on COTS and cloud

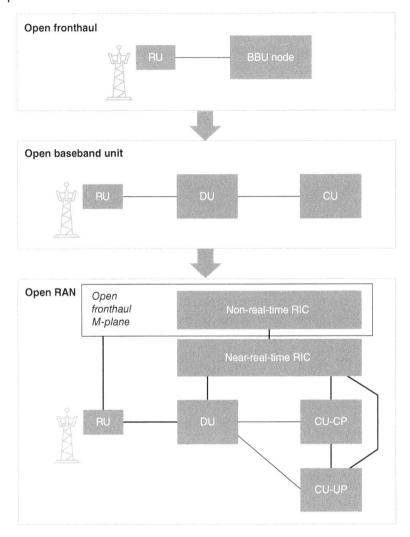

Figure 1.1 The path toward fully Open RAN deployment involves the prior deployment models for open fronthaul and open BBU. The thick lines represent the O-RAN-defined open interfaces.

platforms. This scenario extends the disaggregation of the CU so that it is divided into separate user and control planes. The RIC is essential in this scenario to house real-time analytics, self-optimized network concepts, and radio resource management applications so that the CU and RIC reside in edge cloud.

This list presents the major steps that augment the level of disaggregation and openness. It should be noted that the development of the interfaces for open management of all the relevant components plays an essential role in taking the O-RAN concept work in practice. Observing Figure 1.1, apart from the depicted interfaces, the M-plane of the O-RAN fronthaul thus connects to the O-RAN entities, COTS hardware (O-cloud) as well as the core network segment.

As a result, the Open RAN movement allows new stakeholders to develop focused products through the defined open interfaces. The O-RAN Alliance has made this reality for the respective signaling and data flows and has developed the RAN architecture to be virtual, intelligent, and interoperable.

The first chapters of this book describe the principles of the Open RAN concept, further detailing the respective deployment aspects in Chapter 8.

1.2.5 Security Aspects

The Open RAN concept evolves the interfaces to allow additional ecosystem stakeholders to join the network operation and service provision. The concept is rather novel, so despite the good intentions to open the RAN interface, this also needs to be considered in the evolving fraud and security landscape. Basically, the greater openness might also open up new attack vectors that were not known in previous networks.

As reasoned in [12], Open RAN is affected by many of the similar security risks that are present in previous RAN deployments; furthermore, apart from these common security risks, there are also new or more pronounced ones along with the deployment of Open RAN. These vulnerabilities can relate to, for example, AI/ML and cloud security risks.

The O-RAN technical specifications provided by their latest working group 11 present security requirements, for example, in the O-RAN Security Requirements Specification, in order to mitigate part of these new risks. The evolution of the O-RAN must also include continuous monitoring of the landscape, and in the future, sharing the information on new vulnerabilities among the stakeholders can also play a key role in securing the networks. Chapter 7 discusses the Open RAN security aspects in greater detail.

1.2.6 Commercial Deployments

The initial phase for the deployments has been still rather tranquil, though, in 2023. The 5G Americas presents a summary of the initial deployment landscape that indicates observations include the following the [13]:

- *3G, 2020*: Rakuten introduces O-RAN 5G MIMO RRH upon Intel and NEC solutions.
- *1Q, 2021*: Deutsche Telecom launched O-RAN in 25 cities based on Mavenir 4G and 5G solutions.
- *1Q, 2022*: Vodafone UK launched a commercial 5G O-RAN site that comprises Samsung RAN, COTS hardware of Dell and Intel, and software of Wind River.
- *2Q, 2022*: DISH Network launched a nationwide 5G O-RAN using Mavenir and Samsung cloud RAN, Nokia core that bases on AWS hyper-scaler. Later, DISH Network included virtualized 5G components of Mavenir, Altiostar, and Fujitsu, the cloud platform of VMware.

1.3 Focus and Contents

This Open RAN explained book details the new architectures, functionalities, and interfaces of the Open RAN model and explains the standards, principles, and practical aspects the operators need to know to be prepared for the new era. It also presents the challenges and opportunities of the modularization and introduces scenarios and recommendations for the optimal network deployment, considering techno-economic aspects and new security landscape [14–16].

This book summarizes the latest key features and aspects of the Open RAN to provide the readers with a single source of information that eases network deployment, device designing, and education of personnel and students interested in advanced mobile communications. This book presents up-to-date information on the practical aspects of the Open RAN and serves as a guideline to technical personnel of operators and equipment manufacturers, as well as for telecom students. Previous knowledge of mobile communications and especially 3GPP radio networks would help in capturing the most detailed messages of the book, but the modular structure of the chapters ensures the feasibility of the book also for the readers not yet familiar with the subject.

The readers, who are primarily from mobile equipment engineering, security, network planning, and optimization as well as application development teams, would benefit from the contents of this book as it details some highly novel aspects of the O-RAN. This book tackles the needs of specialists who wish to capture fast the new key aspects of O-RAN and its differences between the previous radio network models and to understand the key aspects for device planning, network deployment, and app development based on the available specifications of the O-RAN and preceding efforts of the 3GPP.

The primary audiences of this book are technical personnel of network operators, network element, and user device manufacturers, as well as telecom professors and students in advanced academic levels and hands-on courses. The book is also beneficial to telecom regulators, standardization bodies such as 3GPP and O-RAN, service providers, consultancy companies, sales, marketing, and solutions selling personnel.

The content of this book is modular and the chapters are independent from each other in order to ease the exploration of the desired topic.

Chapter 2 presents an overview to Standards Development Organizations relevant to Open RAN, covering the key standards of, for example, 3GPP and O-RAN, including modules and interfaces, requirements on Open RAN-related functional and performance aspects, and interoperability considerations.

Chapter 3 outlines the overall cellular communications architecture and provides an overview of different components of the RAN. Specifically, the history of RAN evolution is described with regards to not only generational evolution, but also in the context of performance indicators. This chapter also describes the introduction of new IT technologies in the telecoms, and their impact on 5G RAN evolution are also described. For example, evolution from distributed RAN to virtualized RAN.

Chapter 4 presents the details of the O-RAN architecture, where O-RAN alliance's profile on 3GPP specifications and O-RAN alliance's own enhancements are described. Readers are given holistic view from the traditional RAN nodes to intelligent operation & management and fronthaul.

Chapter 5 describes TIP activities revolving around Open RAN in more detail. It outlines the concept of release in TIP and how the release has evolved. In addition, details on TIP's projects in the area of Open RAN and testing & validation process are provided.

Chapter 6 presents some of the relevant Open RAN use cases and respective methods to assess them. The chapter discusses selected use cases representing the benefits of

Open RAN and their technical realization including traffic steering, Quality of Service, Quality of Experience optimization, massive MIMO, Network Slicing and Service Level Assurance, and V2X Handover. Also, uncrewed aerial vehicle (UAV) resource allocation and applications are discussed.

Chapter 7 discusses the security aspects of Open RAN including the current threat landscape with new types of threats and attack vectors in cloud and virtualized environment and presents some recommendations for protection.

Chapter 8 discusses Open RAN planning and deployment scenarios, RAN dimensioning, and service-based planning aspects. The chapter also presents interoperability and roaming aspects, measurements, optimization, and practical strategies for the selection and preparation of equipment including re-farming of RAN to Open RAN.

References

1 O-RAN Alliance, "O-RAN specifications," O-RAN Alliance, 2022. [Online]. Available: https://www.o-ran.org/specifications. [Accessed 14 December 2022].

2 Open RAN Alliance, "Open RAN Alliance," [Online]. Available: https://www.o-ran.org/. [Accessed 5 October 2023].

3 Telecom Infra Project, "Telecom Infra Project: Who we are," [Online]. Available: https://telecominfraproject.com/who-we-are/. [Accessed 5 October 2023].

4 Telecom Infra Project, "Open RAN MoU Group Page," [Online]. Available: https://telecominfraproject.com/openran/. [Accessed 5 October 2023].

5 Open Networking Foundation, "ONF Mission," ONF, [Online]. Available: https://opennetworking.org/. [Accessed 5 October 2023].

6 Open RAN Policy Coalition, "Mission," [Online]. Available: https://www.openranpolicy .org/. [Accessed 5 October 2023].

7 Mavenir, "Open RAN by the Numbers," Mavenir, [Online]. Available: https://www .mavenir.com/resources/open-ran-by-the-numbers/. [Accessed 5 October 2023].

8 P. Lipscombe, Data Centeer Dynamics, 11 May 2023. [Online]. Available: https://www .datacenterdynamics.com/en/analysis/how-real-is-open-ran/. [Accessed 5 October 2023].

9 S. Miles, "Open RAN equipment market forecast 2023–2030," Electronic Specifier, 31 July 2023. [Online]. Available: https://www.electronicspecifier.com/news/analysis/open-ran-equipment-market-forecast-2023-2030. [Accessed 6 October 2023].

10 TeckNexus, "5G Networks Magazine: Open RAN," TeckNexus, June 2021. [Online]. Available: https://tecknexus.com/5g-network/current-state-of-open-ran-countries-operators-deploying-trialing-open-ran/. [Accessed 6 October 2023].

11 Z. Ghadialy, "Introduction to O-RAN Philosophy," Parallel Wireless, 14 April 2021. [Online]. Available: https://www.parallelwireless.com/blog/introduction-to-o-ran-philosophy/. [Accessed 6 October 2023].

12 NTIA, "Open RAN Security Report," NTIA, May 2023. [Online]. Available: https://www .ntia.gov/sites/default/files/publications/open_ran_security_report_full_report_0.pdf. [Accessed 6 October 2023].

13 C. Pearson, "An Update on Open RAN," 5G Americas, February 2023. [Online]. Available: https://www.5gamericas.org/an-update-on-open-ran/. [Accessed 5 October 2023].

14 O-RAN Working Group 11, "Security Requirements Specifications," O-RAN Alliance, March 2023.

15 Open RAN Policy Coalition, "Open RAN security in 5G," Open RAN Policy Coalition, April 2021. [Online]. Available: https://www.openranpolicy.org/wp-content/uploads/2021/04/Open-RAN-Security-in-5G-4.29.21.pdf. [Accessed 14 December 2022].

16 Fujittsu, "A brief look at O-RAN security," Fujitsu, 2022. [Online]. Available: https://www.fujitsu.com/global/documents/products/network/Whitepaper-A-Brief-Look-at-O-RAN-Security.pdf. [Accessed 14 December 2022].

2

Open RAN: Journey from Concept to Development

2.1 Overview

Mobile communications network is of limited utility if it enables only its subscribers to communicate with one another. The power of mobile communications network comes from following an interoperable standard that enables subscriber from a network operator to communicate with a subscriber from virtually any other network operator. Mobile communications standards are the most fundamental basis for achieving this, and this is no exception for Open RAN. This section provides an overview of how Open RAN standards are created based on the industry stakeholders requirements, the different organizations that develop Open RAN standards along with the collaborative relationships, and how interoperability is ensured in practice.

2.2 Requirements

The telecommunications industry has been endeavoring to upgrade and enhance the quality of connectivity, and this has been a remarkable success in that more and more people in the world are able to enjoy better connectivity as time passes by. This, however, does not come without cost and requires a significant effort in research, development, standardization, commercialization, and operation. Furthermore, with consolidation of players in the saturating telecoms market in the midst of varying requirements across different regions, it is becoming more difficult for the industry to elastically and quickly respond to the various requirements that come not only from the "traditional" consumers but more and more prominently from other sectors of the society. 5G brings even more complexity to requirements, as use cases are expanded beyond traditional personal communications to communications between "things." The ever-increasing demand for data is augmenting the difficulty, and more radio base stations need to be deployed to satisfy the new and more demanding connectivity requirements. This means that the industry is faced with the challenge of improving the performance of the network while reducing its costs of deployment and ownership.

In this context, Open RAN was envisioned to increase competition, foster innovation, and enable new service models. With opening up of previously proprietary interfaces, it will be possible to increase competition in a quite consolidated RAN market, bringing more

Open RAN Explained: The New Era of Radio Networks, First Edition.
Jyrki T. J. Penttinen, Michele Zarri, and Dongwook Kim.
© 2024 John Wiley & Sons Ltd. Published 2024 by John Wiley & Sons Ltd.

innovation and potentially finding more cost-efficient solutions. The Open RAN aims to utilize latest intelligence technologies to enable new use cases of RAN and therefore foster innovation of RAN. Finally, the flexibility of the RAN overall will enable new service models, as deployments are now more flexible and new services can be deployed with less lead time and cost.

More precisely, the Open RAN builds on the four principles of openness, virtualization, intelligence, and programmability.

- *Openness*: the interfaces between different nodes/functions of Open RAN need to be open to allow multi-vendor interoperability and coexistence across functions.
- *Virtualization*: the software should be cloud native running on the commercial-off-the-shelf (COTS) hardware rather than vendor-proprietary hardware.
- *Intelligence*: the latest intelligence technologies such as artificial intelligence (AI) and machine learning (ML) techniques are applied to the RAN, and the optimization of RAN can be done by the third party rather than this being the responsibility of only the hardware vendor or the network function provider.
- *Programmability*: the optimization and the operation of the RAN can be configured and adopted based on the specific deployment scenario and required performance by means of application programming interfaces (API) that allow the operator to configure the network programmatically.

2.3 Standards

2.3.1 3GPP

The most popular and representative mobile communications network standard organization is the third Generation Partnership Project (3GPP). 3GPP is where the technical specifications of mobile technologies (2G GSM, 3G UMTS, 4G LTE, and 5G) are created and maintained. It covers the aspects of the mobile networks ranging from mobile devices (called UE: user equipment), base stations, to core network [1].

As implied in the name partnership project, it should be highlighted that the technical specifications of 3GPP per se, and unlike national or regional standards, do not have any binding power. It is through the transposition of the technical specifications, often consisting of the simple addition of a cover page of the 3GPP's organizational partners (OP), that the specifications become binding. 3GPP's OPs include the SDOs covering all four corners of the world: China (CCSA), Europe (ETSI), India (TSDSI), Japan (TTC, ARIB), Korea (TTA), and North America (ATIS) all transpose technical specifications into standards applicable in their jurisdiction. Thus, 3GPP's technical specification becomes binding, for example, in Europe, when the specification is transposed by ETSI.

3GPP is divided into three major technical specification groups (TSG) depending on the scope and/or the stage of the standardization [2]. TSG Radio Access Network (RAN) covers the RAN-related aspects, meaning that network entity that deals with the control and management of radio resources is under its scope. TSG Service and System Aspects (SA) cover high level or common aspects such as service requirement, architecture, security,

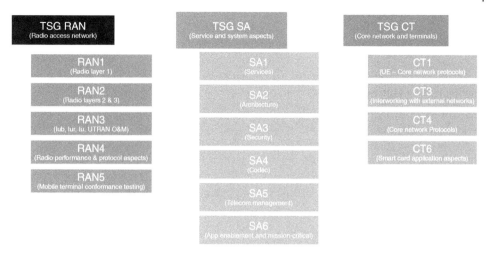

Figure 2.1 3GPP, TSGs, and working groups.

codec, enablement of applications, mission-critical applications, and management of telecom networks. TSG Core Network and Terminals (CT) covers the aspects of the core network and the interface between the user device (UE) and the core network. Each TSG has a number of child groups, which are called the working groups (WGs), and the details can be found in Figure 2.1: 3GPP, TSGs, and WGs.

Out of these groups, those with most relevance to Open RAN are the TSG RAN and its WGs. RAN WG1 (abbreviated as RAN1, similarly for other WGs) deals with the physical layer procedures between the device (UE) and the RAN, for example, what channels and types of multiplexing are used in different generations of radio access technology (RAT). RAN2 also focuses as RAN1 on the "air interface," but looking at different layers of the Open System Interconnection (OSI) model and different protocols. It is in RAN2 where protocols such as Radio Link Control (RLC), Packet Data Convergence Protocol (PDCP), and Service Data Adaptation Protocol (SDAP) are specified. RAN3 covers the interfaces among the nodes of RAN. RAN4 is responsible for the radio frequency (RF) aspects of the RAN, and it is where test requirements and procedures for the RAN nodes (e.g. base stations, repeaters, relays, etc.) are standardized. RAN5 develops the UE conformance test specifications at the radio interface, which covers not only the interface between RAN and the device but also the interfaces between the core network and device which is standardized by CT1. Considering the success of 3GPP specifications and that the system that it defines is a truly global standard with hundreds of mobile operators and billions of devices, it is critical for Open RAN adoption that it maintains a high level (ideally full) compatibility with 3GPP specifications. It is especially important that devices that are already deployed in the network can seamlessly work with Open RAN technologies. As observed in the past, a possible path that organizations developing Open RAN standard could follow is to ensure that 3GPP endorses their work and includes it in their specifications, but as a minimum the organizations working on Open RAN should be mindful that breaking the compatibility with the 3GPP system is going to make it harder to gain a solid foothold in the market. We will describe in further detail in the following subsections some of the organizations that are developing Open RAN standards.

2.3.2 O-RAN Alliance

O-RAN Alliance is an alliance of mobile network operators, vendors, and research and academic institution to promote the opening of the RAN (i.e. Open RAN). That is, it aims to transform the current RAN industry/technology to more intelligent, open, virtualized, and fully interoperable network. O-RAN Alliance envisions a more competitive and vibrant supplier ecosystem of RAN with faster innovation to improve user experience, where the mobile networks will be able to enjoy improvements in the efficiency of RAN deployments and operations. Its board consists of major mobile network operators, indicating the high level of interest from the network operator community on the concept of Open RAN [3] (see Table 2.1).

It is worth noting that out of the 15 board members, five founding mobile network operator members have permanent seat in the board, while other members are elected every two years.

O-RAN Alliance was first formed in February 2018, when its predecessor xRAN Forum (founded in June 2016) merged with C-RAN Alliance, an organization promoting the adoption of Cloud RAN. xRAN Forum had a similar vision as O-RAN Alliance did, where it aimed to promote a software-based, extensible RAN (hence xRAN) and to standardize critical elements of the xRAN architecture [4]. It is interesting to note that xRAN Forum was led by some of the O-RAN Alliance's founding members (AT&T, Deutsche Telekom, and SK Telecom) and Dr. Sachin Katti, professor from Stanford University. Currently, O-RAN Alliance is led by the board that makes important decisions and the Technical

Table 2.1 O-RAN Alliance board members (as of October 2023).

Company	Type
AT&T	Founding Member
China Mobile	Founding Member
Deutsche Telekom	Founding Member
NTT DOCOMO	Founding Member
Orange	Founding Member
Bharti Airtel	Elected Member
DISH Network	Elected Member
KDDI	Elected Member
Rakuten Mobile	Elected Member
Reliance Jio	Elected Member
Singtel	Elected Member
Telefonica	Elected Member
TIM	Elected Member
Verizon	Elected Member
Vodafone	Elected Member

Steering Committee (TSC) that decides or gives guidance on O-RAN technical topics and approves O-RAN specifications prior to the board approval and publication. TSC is currently co-chaired by Dr. Sachin Katti, who led xRAN Forum, and Dr. Chih-Lin I, chief scientist of China Mobile.

Under the governance of the board and the TSC, O-RAN Alliance is further divided into three main groups. First, there are workgroups that build technical specifications for the O-RAN architecture, where the activities are supervised by the TSC. Second, there are software communities to develop and maintain codes for realizing the specifications. Third, there are centers for testing prototypes and products for integration. Therefore, O-RAN Alliance spans wide range of activities from specification documents to codes for implementation and testing for ensuring interoperability (see Figure 2.2).

As of October 2023, the technical specifications of O-RAN are divided among eleven technical workgroups as follows:

- *WG1*: Use Cases and Overall Architecture Workgroup.
- *WG2*: The non-real-time RAN Intelligent Controller (RIC) and A1 Interface Workgroup.
- *WG3*: The Near-real-time RIC and E2 Interface Workgroup.
- *WG4*: The Open Fronthaul Interfaces Workgroup.
- *WG5*: The Open F1/W1/E1/X2/Xn Interface Workgroup.
- *WG6*: The Cloudification and Orchestration Workgroup.
- *WG7*: The White-box Hardware Workgroup.
- *WG8*: Stack Reference Design Workgroup.
- *WG9*: Open X-haul Transport Work Group.
- *WG10*: OAM Work Group.
- *WG11*: Security Work Group.

WG1 covers the O-RAN architecture and use cases, working closely with other workgroups and is responsible for the overall architecture. WG1 consists of three subgroups dedicated to specific tasks: the Architecture task group, the Network Slicing Task Group, which investigates how Network slicing applies to O-RAN defining in the process new use cases and requirements, and the Use Case Task Group, which is fully dedicated to defining new use cases and refining existing ones. In terms of controlling and management of the radio resource management, WG2 is the group working on the intelligent controller for RAN, more specifically WG2 defines the non-real-time management and optimizations. WG3 develops the near real-time RIC that seeks to control and optimize RAN functions by issuing control instructions based on analysis of RAN data and using AI/ML tools. The workgroups that cover the aspect of transport networks are WG4 and WG9. WG4 specifies fronthaul interface so that it can accommodate multi-vendor interoperable RAN components, while WG9 looks into the transport network aspects such as the equipment, physical media of transmission, and control/management. WG5 covers the specifications of the interfaces among RAN nodes (e.g. between DU and CU), including F1, W1, E1, X2, and Xn. While these interfaces are already specified in 3GPP, WG5 proposes enhancements and profiling that aim to improve the interoperability between equipment provided by different vendors that connect through these interfaces. WGs 6, 7, and 8 look to provide reference designs for the virtualized network, from the hardware (WG7), software (WG6), and finally to the orchestrator (WG8) that orchestrates the common commodity

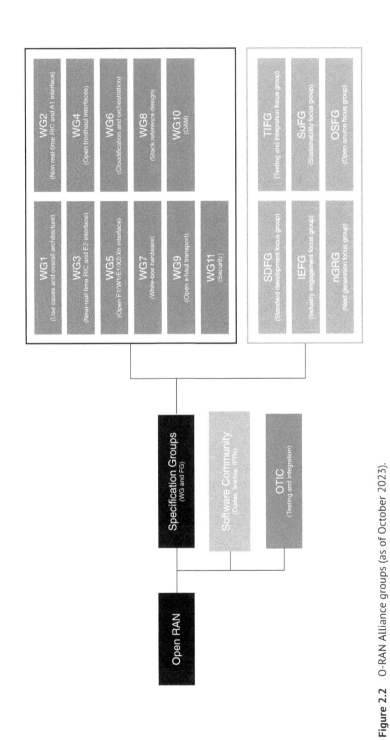

Figure 2.2 O-RAN Alliance groups (as of October 2023).

hardware platform and the software running on top of such platform. WG10 oversees the operation, administration, and maintenance requirements and architecture for O-RAN by specifying the O1 interface that enables the Service Management and Orchestration (SMO) to interact with the other O-RAN components. The latest addition to the O-RAN Alliance organization is the WG11 that focuses on security aspects of the open RAN ecosystem.

In addition to the workgroups, there are focus groups where the scope spans a few WGs or sometimes applies to whole organizations. They are:

- *Standard Development Focus Group (SDFG)* leads standardization strategies of O-RAN Alliance and serve as the main interface to other standard development organizations.
- *Test and Integration Focus Group (TIFG)* plans and coordinates testing and integration aspects of O-RAN Alliance, including the plugfests and third-party Open Test and Integration Centers (OTIC).
- *Open Source Focus Group (OSFG)* is in a dormant mode as its activities have been transferred to O-RAN software community.
- *Industry Engagement Focus Group (IEFG)* promotes the O-RAN technology adoption and innovation, including engagement of the O-RAN ecosystem.
- *Sustainability Focus Group (SuFG)* focuses on the environment-friendly mobile networks by enhancing energy efficiency.
- Finally, *next Generation Research Group (nGRG)* focuses on research and adoption of Open RAN principles in the 6G and future network standards.

O-RAN Alliance also operates O-RAN software community, which was a spin-off of OSFG. This community is a collaboration between O-RAN Alliance and Linux Foundation and aims to align a software reference implementation for the O-RAN Alliance's architecture and specifications. As of March 2022, there are 12 projects under the software community, where three are related to the RIC and its applications, another three on the nodes of the RAN (DU and CU), two on the virtualization of infrastructure (infrastructure and orchestration), and other four on miscellaneous topics such as simulation, documentation, OAM, and integration and test [5].

Building on the specifications and software, O-RAN Alliance also operates open testing and integration centers, which enable different vendors to work collaboratively to test and integrate their products. As Open RAN requires interoperability of products and nodes from different vendors, OTIC is an essential component of O-RAN Alliance. In OTIC, proof of concept (PoC), lab trials, and field trials may take place to test the multi-vendor system's compliance and feasibility. Note that OTIC is a third-party center that operates based on the guidelines set by TIFG and not operated by O-RAN Alliance. As of March 2022, there are four OTICs in Europe (Berlin, Torino, Madrid, and Paris) and two OTICs in Asia (Taoyuan City and Beijing) [6].

To ensure that solutions are compliant with O-RAN specifications and guidelines, O-RAN Alliance also issues certificates and badges based on tests. There are three types of tests that lead to certification or badges [6].

- *Conformance test*: test that verifies compliance of a single device to O-RAN interface or reference design specifications. These tests are as defined by O-RAN conformance test specifications, and certification of conformance is given if the device passes the test.

- *Interoperability test*: test that assesses the interoperability of the pairs of devices that are implemented according to O-RAN interface specifications. The tests are as defined by O-RAN interoperability test specifications and interoperability badge is given if the devices pass the test.
- *End-to-end (E2E) tests*: test that assesses the integration of groups of devices that are implemented according to O-RAN interfaces. The tests are as defined by O-RAN E2E test specifications, and E2E system integration badge is given if the devices pass the test.

2.3.3 Telecom Infra Project

Telecom Infra Project (TIP) is a foundation led by service platform provider Meta (formerly Facebook), global network operator groups (BT, Deutsche Telekom, MTN, Orange, Telefonica, and Vodafone), and vendors (Dell Technologies and Intel). TIP is not only focused on the RAN but also covers the building and deploying of open, disaggregated, and standard-based technology solutions in the field of telecommunication network overall. For example, it not only covers opening of RAN but also transport, core network, and fixed broadband networks [7].

TIP was founded in 2016 with 30-member companies, with Deutsche Telekom, Intel, Meta, Nokia, and SK Telecom leading initially. It may come unusual that social network service provider Meta joined forces with the telecom industry in adopting open approach. However, Meta was always interested in opening up the proprietary systems in its data centers with Open Compute Project (established in 2011 outside Facebook). In addition, the power of Meta came from the number of users of its platform where connectivity is a prerequisite, which makes it for Meta to promote open approach of networks such that more people can be connected and hence use Meta's platform. Therefore, Meta has been a leader in TIP's activities, and its strategy should not be overlooked when analyzing TIP's activities [8].

In terms of governance, TIP is led by the board that makes important decisions and technical committee that drives TIP's technical strategy. As of October 2023, the board consists of nine companies mentioned earlier: BT, Dell Technologies, Deutsche Telekom, Intel, Meta, MTN, Orange, Telefonica, and Vodafone. The Technical Committee includes all board companies with an addition of TIM Italia. Under the board and the technical committee, TIP has various project groups, labs, acceleration centers, and exchanges to drive its activities (see Figure 2.3 for details).

Starting with the project groups at TIP, there are three types. The first type is the product groups that are product oriented and focus on the development of requirements, minimum viable products, white papers, and trial results. The second type is solution groups that are more oriented toward specific deployment cases and attempts to provide E2E solutions for the case concerned. The last type is software groups that develop open-source software to support open network. While there are many groups in each, the directly relevant group for Open RAN is Open RAN, and therefore, this section will describe the Open RAN product group in detail [9].

Open RAN product group focuses on supporting development of multi-vendor interoperable product/solution in RAN domain, where 2G, 3G, 4G, and 5G NR technologies are

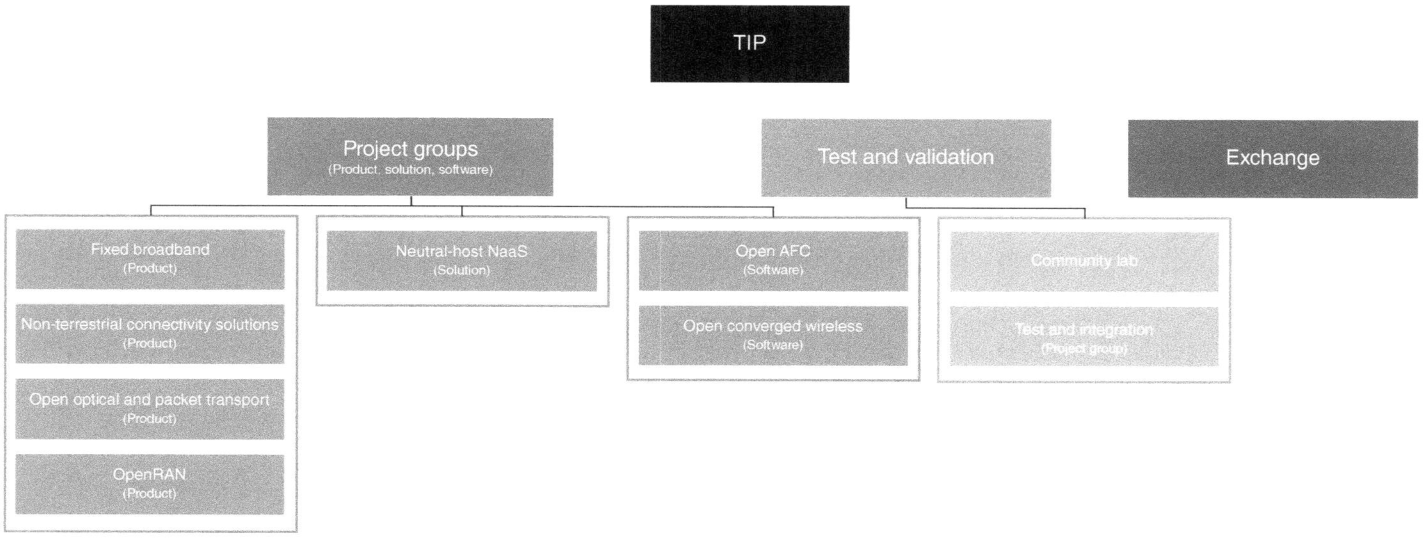

Figure 2.3 TIP organization (as of October 2023).

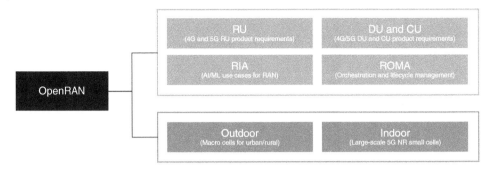

Figure 2.4 Open RAN subgroups (as of October 2023).

all covered. In TIP, Open RAN subgroups are created along two dimensions of components and segments as shown in Figure 2.4. In terms of components, there are four subgroups:

- RU
- DU and CU
- RAN intelligence and automation (RIA)
- ROMA

For segments, there are two subgroups: indoor and outdoor.

In the component dimension, the RU subgroup focuses on defining RU product architecture and requirements. Its goal is to develop white box hardware with open disaggregated architecture such that software from different vendors can perform fully.

The DU and CU subgroup focuses on disaggregating the traditional baseband unit (BBU). Therefore, the subgroup's goal is to develop white box hardware for DU and CU that supports multiple split options in an open interoperable manner.

The RIA subgroup leverages the RIC platform that O-RAN defines and develops AI / ML use cases on top. For example, the subgroup has defined use cases including load, coverage, capacity, interference, massive multiple-input multiple-output (MIMO), and energy management of RAN.

The ROMA subgroup is a relatively novel subgroup that aims to aggregate and harmonize requirements on orchestration and lifecycle management automation for Open RAN.

In the segment dimension, the outdoor macro subgroup facilitates Open RAN deployment by defining requirements, aggregating technology solutions, and developing playbooks for outdoor deployments. It classifies three major use cases: rural, peri-urban, and urban.

On the other hand, the indoor 5G small-cell subgroup aims to address the challenges of large-scale indoor 5G NR small-cell deployments. It does so by defining relevant reference architecture and API/interfaces.

In addition to project groups, there are groups that cover testing and validation aspects of TIP activities. Community labs are third-party labs that enable TIP participant organizations to test and collaborate on disaggregated network solutions. There are three types of TIP community labs with different focus: product labs, integration labs, and deployment labs. In addition, there is a Test and Integration project group that defines test plans, conducts test, and ensures interoperability of the products as specified by other project groups.

Table 2.2 Summary of badges: scope and eligibility.

Badge	Scope	Eligibility
Bronze	Individual network component	Vendor's own lab
Silver	Integrated network layer	TIP community lab/third-party lab
Gold	End-to-end solution	TIP community lab/third-party lab
Operator tested	All types	Tested in operator field trial
Requirements compliant	All types	No test

Exchange is what makes TIP unique. TIP Exchange is where TIP participants can showcase qualified products and solutions, where service providers can easily evaluate technology and establish technology partnerships to realize open networks in practice. TIP Exchange consists of three major components. The first is the marketplace, open marketplace similar to eBay for certified TIP-compliant network equipment products for interested customers to consider. The second is the solutions, where TIP provides integrated solutions of TIP certified products for different use cases. Lastly, there are badges that certifies that a product is interoperable and open according based on TIP activities. These badges provide the foundation for interoperability and is the requirement to be listed on marketplace and/or solutions. There are five types of badges and ribbons that can be obtained (also summarized in Table 2.2) [10]:

- *Bronze badge*: This badge is awarded to products that vendors test in their own labs and requires the approval of the test results by the relevant project group. This badge applies primarily to individual network components and products. It is compulsory for the product to be commercially available, but commercialization can be early stage.
- *Silver badge*: This badge is awarded to products validated in a TIP community lab or an approved third-party lab. The testing is done according to the criteria set by the relevant TIP project group and requires the review by the project group before awarding the badge. This is primarily for integrated network layers, and it is compulsory for products to be commercially available with minimum product support.
- *Gold badge*: This badge is awarded to solutions validated in an E2E environment, where the validation can be done in a TIP community lab or approved third-party lab. The testing is done according to the criteria set by the relevant TIP project group and requires the review by the project group before awarding the badge. This badge is applicable to E2E solutions, integrated network layers, or individual products. It is compulsory that the solutions need to be commercially available with full product support.
- *Operator tested ribbon*: This is awarded to listed products on TIP Exchange that were tested in an operator field trial. Having more operators testing the product means the addition of a number on the ribbon, indicating higher level of maturity. Note that the test must be under commercial traffic conditions. This badge is applicable to individual network components/products, integrated network layers, and E2E solutions.
- *Requirements compliant ribbon*: This is the minimum requirement to get listed on TIP Exchange. This can be obtained when the product is deemed to be compliant with associated project group's requirements. This does not require formal test.

2.3.4 Collaboration of Organizations

The three important organizations in Open RAN are not necessarily mutually exclusive and some of the activities overlap. It is important to map the organizations holistically to identify the relative positions and which organization is relevant for a given company's purpose. To map the organizations, this book employs the stages of commercializing a telecommunications network product/solution. Telecommunication network product/solution starts from standards that define the requirements and functionalities for interoperability. Then, the product/solution is developed based on the standard, which are then tested in the lab. Then, the product/solution will go through trials in the field and even be commercialized in a small scale. Finally, the product/solution is officially commercialized and deployed in the field.

Figure 2.5 outlines the stages and maps the organizations of Open RAN in the relevant stages. 3GPP is the most fundamental organization as it specified standards, the basis of all network product/solution. O-RAN Alliance has little overlap in the standards as it also specifies enhancements to some of the 3GPP-defined interfaces. The main activities of O-RAN Alliance, however, lie in the development of standardized product/solution and testing of it. O-RAN software community provides codes for product/solution development and OTIC tests developed product/solution. TIP has some overlap with O-RAN Alliance activities, as they also define standardized product/solution requirements, operates software groups and testing schemes for product/solutions. Both O-RAN Alliance and TIP issue badges to certify compliance of product/solution under certain conditions. However, TIP is more focused toward commercialization as they promote field trials in the context of E2E solutions for a particular use case. This often results in playbooks for commercial deployments, covering the pre-commercial phase.

The ultimate goal of such collaboration is interoperability. Ensuring interoperability of Open RAN is a continuous process that benefits from regular testing, monitoring, and benchmarking in order to ensure adequate, expected, and continuous network performance in multi-vendor environment.

The most challenging phase is the initial interoperability and integration testing, in which most of the issues can be assumed to take place. These first practical experiences support in the creation of implementation guidelines and possibly lessons learned documentation that help the entire ecosystem adjust the details further. The key is the product and system compliance of expectations, requirements, and feasible level for the interoperability,

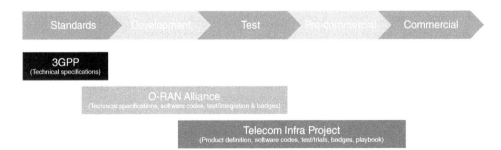

Figure 2.5 Stage of commercialization and how organizations collaborate.

understanding that regardless of the aims to achieve completely seamless experiences, there can be occurrences in practice that require further efforts for integration tuning, such as optional items that are not compatible among different networks. For the expedited Open RAN deployments, adequate level of information sharing among stakeholders, including Open RAN operators and network equipment vendors, plays a pivotal role.

In terms of the specification development, O-RAN Alliance and TIP have created means to test the solutions, e.g. through trials and plugfests on short notice. This expedites the goals achievement considering captured issues and challenges in technical realization and logistics.

2.4 Open Source and Open RAN

The concept of open source is often implicitly implied when discussing the concept of Open RAN. While the two have many similarities and it is the spirit of open source that the Open RAN initiative aims to capture in the industry, the two are not the same and the two concepts should not be used interchangeably. In this context, this section will outline the definition of open source, the role of open source in the telecommunications industry today, myths of open source, and differences between open source and Open RAN.

According to the Linux Foundation [11], open-source software is defined to be a software distributed under a license such that the software is available in source code from without charge or at cost, and the source code may be modified and redistributed without additional permission. Other criteria may apply to its use and redistribution. The open-source initiative provides a more comprehensive definition by providing the criteria for distribution of open-source software as follows [12]:

- *Free redistribution*: the software per se must be free to be redistributed (e.g. no royalty required)
- *Source code*: the distribution must include source code of the software
- *Derived works*: the software must allow derived works and must allow distribution of them in the same open-source license
- *Integrity of the author's source code*: the license must explicitly allow distribution of software built from modified source code
- *No discrimination against persons or groups*: one/group must not be discriminated
- *No discrimination against fields of endeavor*: any field (e.g. research or business) must not be discriminated
- *Distribution of license*: the rights attached to the software must apply to all to whom the software is redistributed
- *License must not be specific to a product*: the rights attached to the program must not depend on the program's being part of a particular software distribution
  - ○ *License must not restrict other software*: it should be possible to distribute parts of a software as open-source software and others as non-open source (i.e. does not have to consist of all open-source software components).
- *License must be technology neutral*: no provision of the license may be predicated on any individual technology or style of interface.

In other words, open-source software is where the source can be accessed free of charge, the software can be modified freely, the derived works are permitted with the same license, and no discrimination is allowed in terms of people/field. From the definitions, it is important to note that the open source is mostly (for definitions, it is all because they only deal with open-source ***software***) in the context of software, and this will be elaborated further in analyzing differences between Open RAN and open source.

The open-source software has been expanding its footprint since the days of 4G LTE, where cloud technologies were adopted to make infrastructure more flexible and scalable (details will be discussed in Chapter 3). Consequently, the open-source organizations that are participating in telecommunications industry are related to cloud. Linux foundation drives cloud native projects and networking projects for the telecoms industry. Openstack focuses on open-source implementation of overall cloud deployments. Open Networking Foundation focuses on achieving programmable networks. Having more open-source organizations to consider means that understanding the telecoms technology requires more in-depth and broader understanding of component technologies. Considering the fragmentations within open source, it will be even more challenging to integrate telecom system.

This leads to the three major myths of open-source software that need to be evaluated. First, most believe that open source will be interoperable and will bring less trouble because it is implementation focused (i.e. action oriented) than requirements focused. While it is true that open source is implementation focused, it does not necessarily mean that the system will be easier to integrate and be interoperable. Rather, open source can be less interoperable because it is implementation focused. Differing views on software can lead to different branches of an open-source code. Second, most consider the open-source software to be free. Indeed, the definition above states that the source code can be accessed free of charge and can be modified freely. However, not everyone is a developer or has sufficient knowledge to access and tailor the open-source software. Therefore, if a particular software or solution is based on open source, it will not necessarily mean that the entire software/solution is open source, and there may be parts or modifications that the user needs to pay for. Especially in the context of enterprises, open source may require a degree of license fee. Lastly, many consider open source to be more secure than proprietary source code because the code is easily accessible, and there will be many developers who will be able to provide peer review. While this is true in many cases, having a large pool of peer reviewers does not necessarily guarantee the quality of the software. The proprietary source code suffers from having a smaller pool of peer reviewers, but it mostly accompanies project manager or developer that takes responsibility for the code. Therefore, in real life, it can be that the proprietary source code will actually go through more thorough review than the open-source code does. An example of this is heartbleed bug in transport layer security (TLS). The bug was unnoticed for over two years after the release (February 2012) and was only reported to the organization in April 2014.

The major difference between open source and standards is that the former is implementation focused and the latter is requirements focused. This is because open source is concerned with working service/component, whereas standards are concerned with interoperability of components and products. Therefore, many distinctions emerge from such difference. First, as standards are focused on interoperability, they tend to define boundaries of components and interfaces, whereas open source does not define boundaries but are concerned with working code. Since interoperability is ensured, standards tend to lead to large adoption when there is commercial value. Open-source technologies, on the other hand, do not necessarily lead to large adoption even if the technology has commercial value. Standards also come with transparency in procedures and records, but open source does not guarantee this. Open source, on the other hand, is agile when it comes to implementing a new idea, but standards tend to take time with firm and rigid procedures.

Where does Open RAN fit in this landscape? It certainly has the spirits of open source, where it tries to focus on the commercial implementation of multi-vendor components and nonproprietary interfaces. However, Open RAN's major focus is on interoperability of the components and interfaces and therefore is not completely open source. It also covers the hardware aspects, which is rarely covered in open-source software. While there are software community to implement what is defined by O-RAN Alliance and TIP, the Open RAN is still heavily dependent on the requirements and the definition of boundaries of interfaces and components. Thus, Open RAN can be interpreted as standards with a flavor of open source.

References

1 3GPP, "About 3GPP Home," [Online]. Available: https://www.3gpp.org/about-3gpp/about-3gpp. [Accessed 13 April 2022].

2 3GPP, "Specifications Groups Home," [Online]. Available: https://www.3gpp.org/specifications-groups/specifications-groups. [Accessed 13 April 2022].

3 O-RAN Alliance, "About O-RAN Alliance," [Online]. Available: https://www.o-ran.org/about. [Accessed 13 April 2022].

4 Businesswire, "xRAN Forum Merges With C-RAN Alliance to Form ORAN Alliance," 17 February 2018. [Online]. Available: https://www.businesswire.com/news/home/20180227005673/en/xRAN-Forum-Merges-With-C-RAN-Alliance-to-Form-ORAN-Alliance. [Accessed 13 April 2022].

5 O-RAN Alliance, "Projects," 2 February 2022. [Online]. Available: https://wiki.o-ran-sc.org/display/ORAN/Projects. [Accessed 13 April 2022].

6 O-RAN Alliance, "Testing & Integration," [Online]. Available: https://www.o-ran.org/testing-integration. [Accessed 13 April 2022].

7 Telecom Infra Project, "OpenRAN," [Online]. Available: https://telecominfraproject.com/openran/. [Accessed 13 April 2022].

8 J. Parikh, "Introducing the Telecom Infra Project," 16 February 2016. [Online]. Available: https://about.fb.com/news/2016/02/introducing-the-telecom-infra-project/. [Accessed 13 April 2022].

9 Project, Telecom Infra, "OpenRAN," [Online]. Available: https://telecominfraproject.com/openran/. [Accessed 13 April 2022].

10 Telecom Infra Project, "TIP Badging," [Online]. Available: https://exchange.telecominfraproject.com/about-exchange/badges. [Accessed 13 April 2022].

11 Linux Foundation, "What Is Open Source Software?," 14 02 2017. [Online]. Available: https://www.linuxfoundation.org/blog/blog/what-is-open-source-software. [Accessed 12 05 2023].

12 Open Source Initiative, "Open Source Definition," [Online]. Available: https://opensource.org/osd/. [Accessed 12 05 2023].

3

Evolution of the RAN

3.1 Architecture of a Mobile Communications System

3.1.1 System-Level Overview

A complete mobile communication system consists of three main components that together enable the transfer of data between two end points where at least one of the end points utilizes wireless technology for part of the communication path.

As illustrated in Figure 3.1, the three parts are the User Equipment (UE), the Radio Access Network (RAN), and the Core Network (CN). The mobile communication system allows both communication that is fully self-contained, between two UEs belonging to the same mobile communication system, and communication between a UE and an external network such as the Public Land Mobile Network (PLMN) or, in more recent systems, an external data network such as the Internet. Two mobile communication systems may also connect to each other allowing a UE served by one mobile communication system to connect to a UE served by another mobile communication system.

3.1.2 User Equipment

The wireless end point of the mobile communication system is referred to as "user equipment" since this is the component of the mobile system that is used by the subscriber to communicate with other subscribers or with a data network. Due to the wireless connectivity, user equipment enables the user to maintain the connection while moving around the network and is battery-operated for this reason.

In the early mobile systems, the main service offered by mobile network operators was the capability of establishing voice communications on the move, and therefore, an early UE was generally referred to as mobile phone as it was functionally equivalent to a fixed phone in terms of capabilities offered to the user. Although early mobile generations started to experiment with data communications such as Short Message Service (SMS), Wireless Application Protocol (WAP), or iMode in Japan, it was only toward the end of the first decade of this century that the user equipment started to evolve and become capable of accessing the many data services that were gaining popularity on the Internet. Better data communication capabilities and the growth of an ecosystem of applications especially

Open RAN Explained: The New Era of Radio Networks, First Edition.
Jyrki T. J. Penttinen, Michele Zarri, and Dongwook Kim.
© 2024 John Wiley & Sons Ltd. Published 2024 by John Wiley & Sons Ltd.

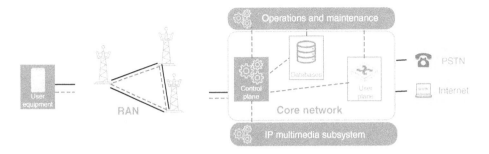

Figure 3.1 System-level architecture UE/RAN/CORE/external networks.

designed to be consumed on mobile devices such as the Apple AppStore and Google Play Store led to the rise of the smartphone: a user equipment that, on top of supporting telephony services, was designed to enable subscribers of the mobile system to enjoy data services traditionally only available from fixed terminals while on the move.

Starting with the fourth generation of mobile system, aptly called 4G, which was widely deployed starting from early 2010, the type and functionality of the user equipment further differentiated in form factor and functionality. Besides devices that are specialized in voice and messaging, now referred to as "feature phones" and smartphones, the more advanced user devices with access to data services, large screens, and sophisticated user interface, there was a growing interest in using the capability of the mobile system also for machine-type communications. This led to the development of user equipment not intended to be used by people but rather by things and that could communicate with each other or with a server. Examples of user equipment supporting machine-type communications include sensors, industrial robots, drones, and, more recently, cars. The collection of these devices forms the Internet of Things (IoT), and it is expected that the number of IoT devices will be in the future an order of magnitude larger than smartphones.

3.1.3 Radio Access Network

The wireless communication between two end points of the mobile system is enabled by the RAN, more commonly referred to as RAN. This part of the mobile system is in fact responsible for transferring the data using radio signals rather than wires, cables, or fiber and, therefore, allowing the user equipment to be reachable while moving.

Conventionally, the transfer of data in the direction of the UE, that is, when the RAN transmits data toward the mobile end point, is referred to as *downlink*. Conversely, the term *uplink* is used to refer to the data that the UE sends to the RAN, that is, in the uplink direction, the RAN plays the role of receiver.

The use of radio waves to transfer information is of course nothing new. For example, satellite television broadcasting uses radio signals to send images to a TV set. Using the terminology defined earlier, we can categorize the communication from the satellite to the TV set as a downlink transmission. It should be noted that this is a type of broadcasting system where the communication is unidirectional meaning that it does not support uplink. An example of a two-way radio communication, that is, a system where each end point can transmit and receive, is the walkie-talkie. Walkie-talkie, on the other hand, is used to communicate from one device to another without the need of a RAN to be deployed. The drawback of the absence of a RAN is that for walkie-talkie communication to be successful, the geographical distance between the transmitting and receiving devices is a limiting factor. The revolutionary aspect of the mobile system RAN is that, unlike broadcasting systems, it supports both uplink and downlink, and unlike walkie-talkies, there is no need for the two users of the mobile system to establish a direct connection with each other in order to communicate, and it is instead sufficient for each end device to be able to communicate with one node of the mobile RAN.

3.1.4 Core Network

The third element that constitutes the mobile system is the core network. The core network connects all the nodes of the RAN with each other and is responsible for the management of the communication between the user equipment and external networks such as the Internet, the public telephony network, or another mobile network. The functionality of the core network in a mobile communications system can be subdivided into session management and mobility management. Session management covers the operations for maintaining the user subscription information up to date, allocating resources needed for the service delivery, authenticating users, securing their communications, and gathering information for charging. Mobility management is a set of procedures that ensure users remain connected as they roam from one part of the RAN to another. The core network also interacts with the Operations and Maintenance System (OAM) as well as, optionally, with the IP Multimedia Subsystem (IMS), which is specialized in providing user equipment services that rely on the Internet Protocol to work.

3.2 Components and Structure of the RAN

3.2.1 The Radio Base Station

The RAN consists of a set of nodes known as radio base station in early mobile systems and NodeB in more recent generations of the RAN. More specifically, NodeB is the name of a 3G radio base station, eNodeB the name of a 4G base station, and gNodeB the name of a 5G radio base station.

A radio base station is, in essence, a system capable of transmitting and receiving radio signals over a given range of frequencies using one or more radio access technologies. Initially, radio base stations were not directly connected to each other, so coordination of operations was delegated to the core network or, in the case of the third-generation mobile system, to a radio control node that could manage a group of base stations. In 4G and 5G groups, base stations are instead connected with a standardized interface allowing the RAN to distribute traffic across the base stations within the same group and manage the mobility without requiring the intervention of the core network.

Each radio base station consists of an antenna system, which is generally constituted by a mast, antennas, and the Remote Radio Head (RRH) and by a Baseband Unit (BBU). Please note that the RRH is also referred in literature as Remote Radio Unit (RRU).

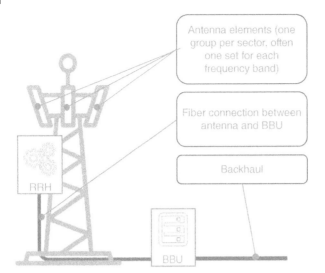

Figure 3.2 High-level representation of a traditional base station (RRH + BBU).

The RRH is responsible for processing the radio signals transmitted or received from the antenna system. In the uplink direction, that is, for radio signals received by the base station, the role of the radio unit is to convert the analog signals to a format that can be more easily processed using digital signal processing techniques; in the downlink direction, that is, for radio signals the base station needs to transmit to the user equipment, the radio unit converts the digital signal into a suitable input for the antenna system.

The BBU role is to prepare the data stream coming from the network for being transmitted to the mobile station in the downlink direction and convert the radio signals received over the air and digitalized by the RRH into a suitable data stream. The connection between the BBU and the core network is known as backhaul, while the BBU and radio unit interface is called fronthaul (Figure 3.2).

3.2.2 RAN Service Area

A radio base station is usually equipped with one or multiple antenna systems designed to transmit and receive radio signals over a given geographical region. The region served by the antenna system, that is, the region where the transmission from the base station can be successfully decoded in the UE and where the UE can successfully establish a communication with the base station, determines the *radio coverage* that the particular base station provides.

In its most common configuration, especially in the case of radio base stations serving sub-urban and rural areas, a radio base station is furnished with three distinct antenna systems, each creating a coverage area that can be modeled as a hexagonal shape, which is known as "cell." Although the practical coverage area can differ considerably from the theoretical presentation of the hexagonal shape, it provides a functional way to approximate the received radio frequency levels and works for the nominal radio link budget dimensioning well enough. As shown in Figure 3.3, by appropriately deploying radio base stations across

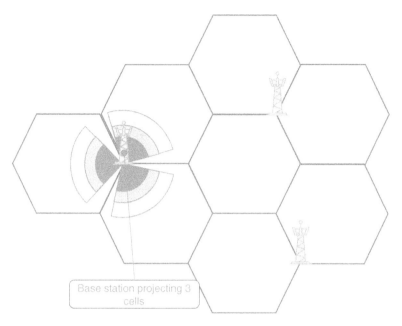

Base station projecting 3 cells

Figure 3.3 Cellular network representation.

the desired service area, the cells form a hexagonal lattice pattern that optimally covers a given portion of territory (service area).

In real world, the shape and size of the service area depend on several factors, and the quality of the service is not uniform, and particularly it degrades further away from the antenna the UE moves. While theoretically a GSM (2G) radio base station could communicate with a device up to 35 km away (in fact, a technique doubling such range to 70 km was also standardized for use in special conditions), in reality, due to the presence of obstacles in the path the radio waves follow between the transmitter and the receiver, such as buildings and trees, the typical size of a cell nowadays is less than 2 km in urban areas and up to 20 km in rural areas, where the lower number of obstacles creates a more favorable environment for the propagation of radio signals.

As the RAN is generally the most expensive part of a mobile system, it pays to discuss the main factors that influence the actual size of each cell in a mobile system. We will also see in subsequent chapters how Open RAN is attempting to address some of these limitations:

- *The nature of the terrain*: Establishing a successful communication requires a minimum signal strength to reach the receiving antenna. Besides the regular attenuation that occurs as the distance between transmitter and receiver increases, obstacles such as buildings greatly contribute in reducing the quality of the link.
- *Frequency spectrum*: As the radio signal power decays proportionally to the square of the frequency used, that is, the higher the frequency used, the bigger is the decay as the signal travels toward the receiver. For this reason, early-generation mobile systems were deployed in spectrum below 1 GHz where a good compromise could be struck between favorable propagation characteristics (coverage) and bandwidth (cell throughput).

- *Mobile station maximum transmit power*: While it would be theoretically possible to overcome the radio signal attenuation by increasing the transmit power, this is unpractical for mobile stations that are generally battery-operated and that are designed to comply to the stringent emission limits set by the World Health Organization.
- *Interference considerations*: Increasing the power used by a radio base station causes the radio signal to "spill over" into the neighboring cells. This radio signal provokes unwanted interference to the communication in these cells and may result in rendering the edge of the cells, where the wanted signal is already weak due to the distance from the transmitter, unusable.
- *Amount of traffic that needs to be supported*: The amount of traffic that can be supported depends on the quantity and quality of the spectrum available as well as on the sophistication of the digital signal processing used to convert radio signals into data and vice versa. A good indication of how well the spectrum is used is provided by the spectral efficiency, which measures how much information can be sent in each Hertz of spectrum in each second. Figure 3.4 presents an example of typical spectral efficiency figures for different mobile communication generations.

The amount of available spectrum and the spectral efficiency of the mobile system result in an upper bound for the traffic that can be supported by a radio base station. If the expected peak traffic is higher than this limit, then it becomes necessary to break up the cell in smaller cells and install additional radio base stations increasing the cost and complexity of the RAN. As each radio base station is responsible for the management of the traffic in the cell or cells it serves, ideally each should be designed to be able to cope with the traffic peak. Nevertheless, there will be instances when the network may not be able to fully accommodate all the traffic (i.e. some of the traffic is blocked), so MNOs typically dimension the network instead to cater for a high proportion of the maximum expected traffic. Figure 3.5 illustrates a typical traffic pattern for a residential and business area and shows that the load

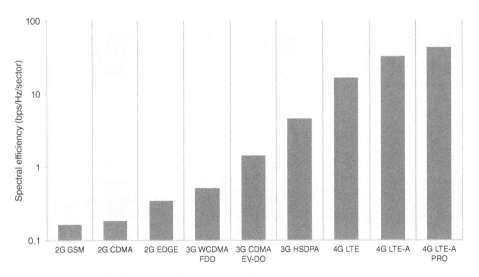

Figure 3.4 Spectral efficiency of different generations of mobile networks. Source: Data taken GSMA, Legacy mobile network rationalisation [1].

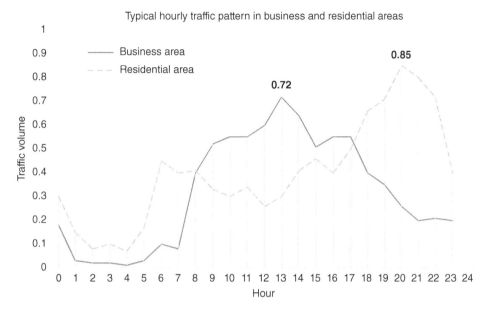

Figure 3.5 Typical traffic pattern in a 24-hour period.

of the base station varies considerably during the course of the day. As a consequence, the resources deployed to manage the highest traffic volume would be unutilized for most of the time, resulting in great inefficiency. When presenting the Open RAN architecture, we will see how this problem of optimizing the resources dynamically has been successfully addressed.

A mobile station moving across the service area from time to time will enter a cell that is provided by a different base station; in this case, the mobile core network will interact with the RAN to ensure that the communication between the mobile device and the network is uninterrupted. The mobile station monitors the received signal strength of the serving radio base station as well as that of neighboring radio base stations. When it detects that a neighboring radio base station signal is stronger, it requests the network to connect to the base station with the stronger signal and to redirect the data stream destined to the mobile station to the new radio base station. This process is known as handover (HO). As the size of the cells becomes smaller, as it is the case for denser networks, HOs become more frequent putting additional strain on the resources in the radio base station. Again, O-RAN architecture provides a means to optimally utilize the radio base station resources.

3.3 RAN Enhancements from Early Mobile System to 4G

3.3.1 Introduction

The RAN has gone through various enhancements from its early days to today. Different generations have adopted proven and reliable technologies at the time of standardization to provide carrier-grade wireless access to their users. To evaluate the enhancements, however, the relevant metrics to measure the quality of a RAN need to be defined in order to be able to benchmark the different generations and make meaningful comparisons. This section describes relevant performance indicators for RAN and describes the RAN enhancements introduced in 4G and how these enhancements contributed to attaining better performances compared to the RAN of the earlier generations. Furthermore, some of the enhancements introduced in 4G are the cornerstone for many of the Open RAN, most significant advancements.

3.3.2 RAN Key Performance Indicators

The evolution of the mobile system has brought about dramatical evolution of the capabilities of the RAN. For example, speed, reliability, latency as well as the amount of traffic that the system as a whole can support. The quality of a mobile system RAN can be measured to a good extent using three key performance indicators (KPIs): the maximum speed at which the RAN can exchange information with the user equipment (throughput); how efficiently the RAN uses spectrum to transfer data (capacity), and how wide the area that the RAN can serve is (coverage). Of course, it is not always possible to have the best of all three and any deployment/generation would need to strike a balance among the three. Nevertheless, all three KPIs have evolved dramatically thanks to the technology evolution in the past 40 years. Some factors that made such evolution possible include:

- *Research in the telecommunications field:* Academia as well as industrial R&D resulted in the discovery of novel strategies to overcome the challenges of radio communications.
- *Advancements to computer technology*: With compute power doubling every two years, some of the digital signal processing strategies that could not be used because of the high computational demands or power, especially in the user equipment, became accessible.
- *Advancements in materials and fixed network connectivity*: Miniaturization of chips, new types of batteries, and more complex antenna systems, all contributed to improving the quality of the connection over the radio interface. Faster connection between RAN and core also enabled RAN engineers to push the boundary of what was possible.
- *Experience from early deployments*: With the first RAN of a mobile system deployed in the 1980s, there is now a vast pool of data on how mobile system performs in real conditions in terms of propagation of radio signals, interference between elements of the RAN, performance at the edge of the network, and so on. This knowledge has allowed the RAN engineers to focus on addressing practical issues and improving the overall performance.

As well as the technology-related, as discussed in Section 3.2.2, a decisive factor in the improvement of the performance of the RAN has been the availability of more spectrum bands for transmitting radio signals.

3.3.3 Early Mobile System Radio Access Networks

The mobile networks, since the very first generation, have evolved in terms of functionality, architectural model, security, and performance in such an extent that it has been justified to distinguish between these different generations. As for the radio interface, each generation has renewed the foundations of the RAN. Whereas the 1G systems were based on frequency-modulated analog radio transmission through Frequency Division Multiple Access (FDMA), 2G introduced Time Division Multiple Access (TDMA) and Code Division Multiple Access (CDMA), of which examples were GSM and IS-95 CDMA systems. The 3G again enhanced the spectral efficiency by introducing Wideband CDMA for the radio access, and 4G is based on more interference-tolerant Orthogonal Frequency Division Multiplex (OFDM).

The most significant benefit of this evolution for consumers' perspective is the augmented data speed and system capacity. Roughly generalized, each new generation has been able to provide about 10 times higher bit rates.

The "traditional" mobile communication systems are based on a set of network elements (hardware and software) to handle their dedicated tasks, such as radio resource management and subscription management. These elements comply with the specified protocol structures for the interoperable signaling and user data flows. The network architectural models have evolved at some extent that can be seen, for example, on the flat architecture of the LTE with integrated radio control functions within the LTE radio base stations.

Nevertheless, even though the aim of the specifications is to provide good interoperability between the network components, in practice, there have been important limitations that have resulted in, for example, issues in interconnecting different RAN providers under the same core network element. While at the time of deployment, the 2G and 3G RAN represented the pinnacle of what the technology and digital signal processing could achieve within the constraints of having to create mass market and, therefore, reasonably expensive products, the RAN designers were fully aware of the weaknesses of the system, for example, coverage and capacity at the cell edge, not very advanced modulation schemes and forward error correction algorithms, interference cancelation strategies, and coordination across different base stations.

3.3.4 The Evolved Universal Terrestrial Radio Access Network

The 4th generation mobile system, which started to be deployed toward the end of the first decade of this century, included several important innovations in the architecture of the RAN that took into account the lessons learned from the experience with 2G and 3G. Also, the fact that 4G is the first system only comprising a packet-switched domain allowed a redesign and simplification of the architecture, especially the architecture of the core network. The 4G RAN, called Evolved Universal Terrestrial Radio Access Network (E-UTRAN), still consists of radio base stations that are called eNodeB, or eNB for short, but the additional controller present in the 3G architecture was removed. As in previous generations (see Figure 3.1), the primary role of the E-UTRAN is to connect a mobile device (known as user equipment or UE) to the mobile system using radio signals. The technology used over the air link is also known as "LTE" (Long-Term Evolution). The E-UTRAN also connects to the Evolved Packet Core (EPC) to enable mobile devices to exchange data with external data networks (e.g. the Internet) or with other mobile devices (Figure 3.6). More specifically, the eNB communicates with the MME for managing the mobility and the session (control plane) and the Serving Gateway (SGW) for the forwarding of the user data (user plane) [2].

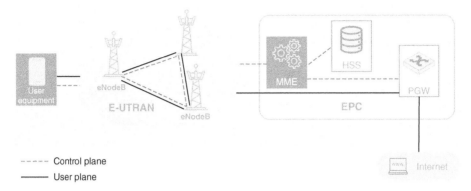

Figure 3.6 4G mobile system architecture.

4G is to date the most successful mobile system, deployed globally by hundreds of mobile network operators with billions of devices not limited to smartphones, but also including sensors, drones, wearables, and so on. Much of the success of 4G is due to some significant advancements of the E-UTRAN described in the following sections.

3.3.4.1 Baseband Unit and Remote Radio Head Split

The 4G radio base station, though logically represented as a single element, splits 4G into two distinct elements performing different functions: the RRH and the BBU.

The RRH contains the base station's components operating on the radio signals and can be further subdivided into transmitter (downlink) and receiver (uplink) sections. The transmitter section of the RRH prepares the data destined to the mobile station to be radiated by the antennas. This preparation involves converting the digital signal into an analog signal at radio frequency and amplifying it to the desired level. The receiver part of the RRH conversely first creates a digital representation of the radio signals impinging the antenna system (analog to digital converter) and then demodulates to move it from high radio frequency to the baseband version that can be processed by the BBU.

The RRH connects to the BBU using the Common Public Radio Interface (CPRI) described in the next section.

The BBU also consists of two distinct pipelines: one for processing the data coming from the core network and destined to the UEs that are served by the eNB and preparing the combined streams to be transmitted by the RRH, and another for processing the raw data coming from the RRH, extracting the individual payload each UE transmitted, and sending it to the core network. The BBU performs a number of operations to optimize the transmission and mitigate the effects of the noise in the radio channel minimizing the error rate. In early E-UTRAN configurations, the BBU is typically located in proximity of the RRH and is connected to the 4G core network with fiber optic cables.

3.3.4.2 Fronthaul Protocol Evolution

With the separation of the RRH and BBU, a mechanism was needed to transfer data produced in the RRH to the BBU and vice versa. In 2003, a consortium of the most prominent radio equipment (RE) manufacturers at the time released the first version of a protocol

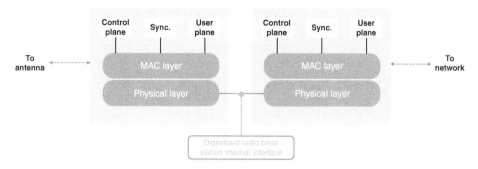

Figure 3.7 CPRI architecture [3].

called CPRI which quickly became the de facto standard in the industry for 2G, 3G as well as early 4G base stations (Figure 3.7).

The CPRI protocol specifies the interface between the RE and the radio equipment controller (REC) defining the means to transport both user data and control data, as well as mechanisms to keep the link synchronization. While not mandated by the standards, it is common practice for the RRH to take the role of RE and the REC to take the role of BBU. We will see in the subsequent chapter how this split corresponds to one of the possible optional splits of the base station defined by 3GPP.

CPRI supports point-to-point topology, where each REC connects to one and only one RE as well as point to multipoint where one REC can connect to several REs. CPRI adopts Time Division Multiplexing (TDM) for the transport of data operating at a constant bit rate. The choice of transferring samples of the antenna signal without any packetization makes CPRI rather inefficient in terms of bandwidth utilization and requires the REC and RE to be located in close proximity. As traffic and frequency bands grow, it becomes necessary to use optical fiber for the interface between REC and RE to increase the costs of the transport. For reference, a 4G base station with three sectors and multiple antenna systems may generate up to 10 Gbps of traffic. Due to constraints in maintaining synchronization, the maximum length of the fiber connection is less than 15 km creating constraints on the topology of the network. CPRI also suffers from the problem that the required bandwidth increases as the number of antennas in the antenna system increases: as 4G may use four or even eight antennas per sector, it may not be possible to meet the total bandwidth requirements using CPRI or its costs could be too high relative to the benefits of separating the RRH and BBU.

At the time when CPRI was first specified, most of the vendors were providing both RRH and BBU, several proprietary additions were made to the baseline standard, and as a result, while theoretically possible, combining RRH and BBU from different vendors required extensive interoperability testing. It has been common for BBU vendors to establish close cooperation with selected RRH vendors possibly reducing the drive to make CPRI more interoperable and increasing the number of proprietary extensions. Along with this evolution, updated testing is required [4].

3.3.4.3 X2 Interface Between Base Stations

The most significant change in the E-UTRAN architecture, compared to previous generations, was the introduction of the X2 interface connecting two eNodeBs. This interface supports the same protocol stack as the S1 interface between RAN and core network. More specifically, the X2 control plane consisting of stream control transmission protocol (SCTP) on top of IP utilizes X2 Application Protocol (X2AP) to enable the exchange of traffic between two eNodeBs. The traffic is tunneled by the X2 user plane which as the corresponding S1 interface stacks GPRS Transfer Protocol (GTP-U) over UDP over IP. With the deployment of X2 between eNodeBs, it becomes possible to forward the user plane between two eNodeBs without the need of involvement from the core network. This direct communication between eNodeBs is especially useful for managing mobility events (e.g. when the UE moves from the coverage of one eNodeB to the coverage of another adjacent eNodeB) but also be able to use two or more eNodeBs to transfer data. As the user plane data for the target UE can be forwarded to the new target eNodeB, there is no need for the core network to redirect the traffic reducing the latency and saving resources in the core.

Furthermore, by exchanging information over X2, the eNodeBs can achieve better coordination, reducing interference, improving the performance at cell edge, and enabling load sharing (Figure 3.8).

It should be noted that although the X2 interface is fully specified in 3GPP, its use in practical deployments is limited to the scenario where the two eNodeBs are provided by the same vendor. This is one of the issues that was addressed by O-RAN Alliance as we will see in the following chapter.

3.3.4.4 User Plane Control Plane Separation

The introduction of advanced smartphones, the growth of video contents, and the availability of multimedia applications resulted in a rapid increase of the user data

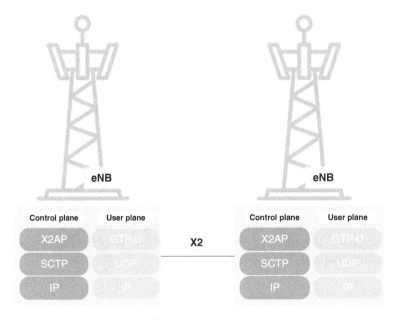

Figure 3.8 Protocol stack of X2 interface.

traffic that the 4G network was expected to support. At the same time, the control plane traffic was unaffected by this growth, and as a consequence, mobile operators started to look at introducing a more efficient core network solution for this traffic pattern.

Borrowing from concepts developed in the context of Software-Defined Networking (SDN), 3GPP completed the specification of the control and user plane separation (CUPS) feature in late 2017. As the name of the feature suggests, CUPS allows to split the functionality related to managing the routing, characteristics, and policies that apply to data traffic (control plane) from the functionality required to actually forward the data traffic to the intended destination (user plane).

Having separate network components managing the user plane and control plane allows operators a more efficient management of the network and unlocks further optimizations. The network nodes that are impacted by CUPS are the SGW, the Packet Gateway (PGW), and the Traffic Detection Function (TDF), which have been each disaggregated into two parts: a network node responsible for the control plane (SGW-C, PGW-C, and TDF-C, respectively) and a network node responsible for the user plane operations (SGW-U, PGW-U, and TDF-U, respectively). It should be noted that this separation did not have any impact on the functional logic of the system: as we can see in Figure 3.9, new interfaces have been defined between the control plane node and the user plane node (Sxa, Sxb, and Sxc) and, where required, interfaces that were carrying both user plane and control plane data have been split.

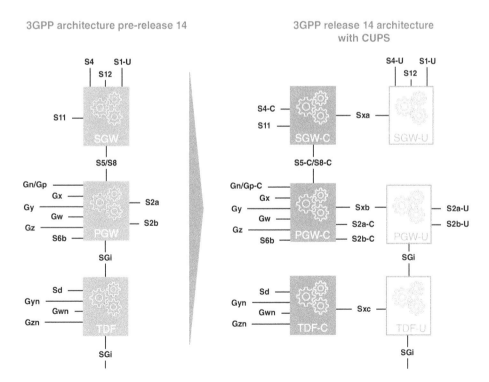

Figure 3.9 Split of nodes and new interfaces from traditional architecture to CUPS. Source: Reproduced from [5].

The main advantage of CUPS is in the fact that the user plane nodes and control plane nodes can be scaled independently of each other greatly reducing the share of unused resources, especially control plane resources, that is typical of integrated network nodes. With CUPS, as the traffic grows, more user plane nodes can be deployed and connected to a pool of control nodes that can be centralized in a data center or distributed. CUPS also allows for independent evolution of the user plane and control plane functionality; furthermore, as the control plane can select the most appropriate user plane function to deliver the traffic to the user, there are benefits in terms of efficiency and latency.

Although CUPS is an evolution of the system that is transparent to the RAN, its importance should not be underestimated and it may be argued that the 5G RAN architecture would have looked quite different if CUPS was not specified. CUPS signaled the intent of the mobile standardization community to adopt technologies that were first developed to address IT systems and could be seen not only as a further step toward the convergence of fixed and mobile networks but also as a first step toward the full separation of user plane and control plane across the whole mobile system, including the RAN with 5G.

3.4 Role of Information Technology in the Evolution of the RAN in 5G and Open RAN

3.4.1 Introduction

Before looking at the evolution of the RAN from 4G to 5G RAN and to O-RAN, it is critical to become familiar with some of the technical enablers developed in the information technology (IT) world that have been adopted and repurposed to enhance the mobile system.

Advances in information technology have been driving the architecture of computers and electronic equipment to a more flexible and disaggregated architecture. While these advancements were initially confined to the realms of IT, in recent years, also telecommunications started to take advantage of them, with CUPS described in Section 3.3.4.4 being just an example of such a phenomenon.

This section describes the technical enablers from IT that have been in part adopted by the 5G RAN and that enable the O-RAN technology, including virtualization, cloudification, and SDN [6, 7].

3.4.2 Virtualization

Virtualization refers to "the simulation of the software and/ or hardware upon which other software runs" or to "the use of an abstraction layer to simulate computing hardware so that multiple operating systems can run on a single computer" [8]. For example, virtualization enables one to operate a classic Minesweeper game by emulating Windows 95 to run on your personal computer which may be using a different operating system. In the context of telecommunications network, the virtualization technology is applied such that there is a separation of software and hardware layers, as shown in Figure 3.10. Whereas the traditional network equipment came in a black box with software and hardware bundled together, virtualization allows the software network function to run on a pool of common

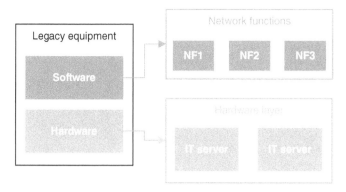

Figure 3.10 The concept of network function virtualization.

off-the-shelf hardware resources. This is called Network Function Virtualization (NFV) and is indeed defined as "principle of separating network functions from the hardware they run on by using virtualization techniques" [9–11].

With the separation of hardware and software layers, NFV can potentially offer the following benefits [12]:

- Reduced equipment cost by exploiting the economies of scale of common off-the-shelf hardware.
- Reduced time to market and accelerated innovation cycle by decoupling software development from hardware investment, enabling network function development and deployment in a more rapid cycle.
- Increased efficiency of network equipment by enabling multi-version and multi-tenancy, where resources can be shared across services and customer bases.
- Creation of open ecosystem, where new entrants and small players can now enter the market of software/hardware.

The main driver for transporting the concepts of NFV in the telecommunications world is the European Telecommunications Standards Institute (ETSI) Industry Specification Group (ISG) NFV. NFV architecture is defined in [13], where the software, Virtualized Network Function (VNF), is decoupled from the hardware, NFV Infrastructure (NFVI). Both are managed by the NFV management and orchestration. The detailed architecture is shown in Figure 3.11.

The architecture consists of the following nodes:

- VNF is a software implementation of network functions that were previously served in a dedicated physical equipment.
- EM stands for Element Management and typically manages one or several VNFs.
- NFVI stands for NFV Infrastructure and consists of the hardware resources (computing, storage, and network) along with the software that abstracts the resources, making the infrastructure seem like a single entity in the VNF perspective.
- Virtualized infrastructure manager manages the infrastructure resources and the interaction of VNFs with those resources.
- VNF manager manages the life cycle of VNFs (e.g. instantiation, update, and query).

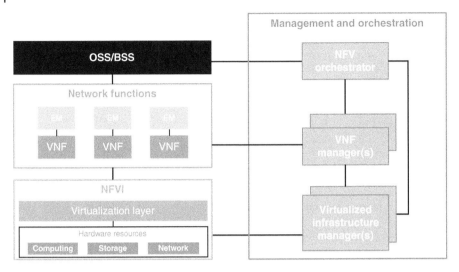

Figure 3.11 NFV architecture. Source: Adapted from [13].

- NFV orchestrator is in charge of the orchestration and management of the NFV infrastructure and software resources, realizing the network services on NFVI.
- OSS/BSS stands for Operation and Business Support Systems

NFV has been optional in the architecture of 4G networks, especially the RAN. However, 5G networks are designed under the assumption that the network functions are virtualized. In the context of RAN, while 4G enabled the split between the RRH and the BBU, 5G enables more refined separation and deployment of the BBU by separating them into Distributed Unit (DU) and Centralized Unit (CU). Therefore, NFV becomes even more relevant for the RAN in the 5G era.

3.4.3 Cloudification

Cloudification refers to leveraging the benefits of cloud computing in providing services, where cloud computing is defined by [14] as "a model for enabling ubiquitous, convenient, on-demand network access to a shared pool of configurable computing resources that can be rapidly provisioned and released with minimal management effort or service provider interactions." This means that cloudification requires leveraging infrastructure where resources can be shared and configured with minimal interaction and management effort and requires the service user to connect to those resources via connectivity [12]. The concept can be simplified as shown in Figure 3.12, where local servers that used to provide the service with dedicated resources are centralized in the central server farm, with resources being shared and pooled among various services (note: the concept of edge will be covered later in this subsection).

By doing so, five benefits can be realized (rephrased and reworded from [14]):

- *On-demand self-service*: a user/consumer can utilize the computing resources of the cloud as needed without having to interact with personnel of the service provider.
- *Support for heterogeneous environment*: use of connectivity enables access to services via different types of devices (e.g. smartphone, tablet, and computers).

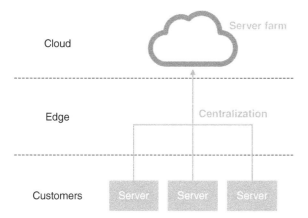

Figure 3.12 Simplification of the cloudification concept.

- *Resource pooling*: the provider's computing resources are pooled to serve multiple consumers. This means that the user/consumer does not know the exact location of the computing resources he/she uses but can abstractly identify the location at high level (country, data center, etc.).
- *Scalability*: resources and capabilities can be elastically provisioned and released, where they can be scaled out depending on the demand.
- *Measurement and transparency*: cloud systems can control and optimize resource automatically by metering their resources and indicators. With this, the resources can be monitored, controlled, and reported, providing transparency for both the provider and the user/consumer.

The centralization, however, also comes with drawbacks. For example, the latency perceived by the users is larger than when the service is provided locally. Furthermore, there can be security and privacy issues where the user/consumer does not want the data/service to reside outside of its premises. To resolve these issues while partly leveraging the benefits of cloud computing, the concept of edge cloud is introduced as shown in Figure 3.13, where some of the functionalities of the cloud are deployed at the *edge* of the network closer to the customers (devices), providing more responsive services in more secure manner.

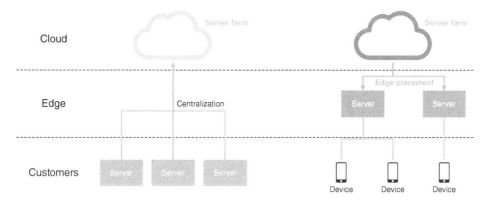

Figure 3.13 Cloudification vs. edge computing.

Indeed, *edge* is anything that is not a "data center cloud" [15], and any attempts to place data center cloud functionalities closer to the customers can be considered edge computing. Specifically in the context of mobile networks, edge computing takes place at the edge of the network, where the RAN hosts some of the core network and cloud functionalities to be used by the customers in proximity. This not only brings benefits to the network and the users but also enables the possibility to use computing resources and functionalities placed in the edge (RAN) as a platform to serve different types of services and adjacent industries (i.e. verticals) [16].

3.4.4 Software-defined Networking

SDN refers to "the physical separation of the network control plane from the forwarding plane, and where a control plane controls several devices" [17], where the control plane refers to the layer that makes decision on where to forward the data and the forwarding plane (also called data plane) refers to the layer that actually forward the data. The traditional network equipment had both layers, having the decisions on where to forward data optimized locally. SDN, on the other hand, attempts to centralize the control plane such that the optimization can be done on a global (or semi-global) manner. This change is depicted in Figure 3.14.

With centralization of the control plane and disaggregation of the layers, the SDN brings the following benefits (rephrased from [17]):

- The control of the network can be directly programmed, allowing the network to be adapted to the requirements of the customer/user.
- The network becomes more agile, as the control plane can dynamically adjust to network-wide traffic flow.
- The network management can be quickly done by network managers themselves with automated SDN programs they write, without having to depend on proprietary software.

In short, SDN allows the global view of network to be obtained by the control plane, allowing global optimization of the network and customers to directly program the

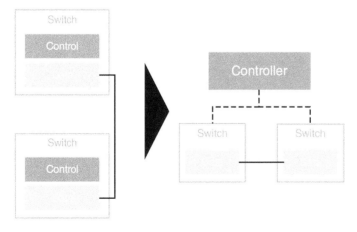

Figure 3.14 Simplified concept of SDN.

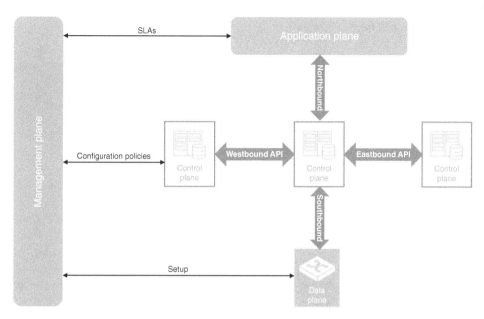

Figure 3.15 Generic architecture of SDN. Source: Adapted from [18].

configuration/management of the network. The latter should be emphasized in that the network now becomes something that can adapt to the customer's needs rather than the customer adapting to the network. Therefore, the architecture of SDN is complete when the customers come into the ecosystem and are able to configure/program the network themselves, as shown in Figure 3.15.

In the generic architecture, there are two more planes: management and application. The management plane is required to ensure configured policies are enforced in the control plane and to set the data plane entities. The application plane enables the users/consumers to program and configure the network, where the Service Level Agreements (SLAs) are fed into the management plane to ensure the agreement is enforced across other layers. The interface by which the control plane entity interacts with other entities depends on which layer it interacts with. When the interface is within the control plane (i.e. between controllers), the interface is an East/Westbound Application Programming Interface (API). Southbound API is when the control plane entity interacts with data plane entities, while Northbound API is when the control plane entity interacts with application plane entities.

In the mobile network context, SDN is realized in various forms. First, 5G networks are designed under the cloud-native principle, which means that it comes under the assumption that the core network is cloudified, with network functions virtualized and two layers (control and data planes) decoupled. Second, the separation of control and data plane is realized in 4G systems as CUPS, where the user plane refers to the data plane. CUPS enables 4G network to adopt SDN by disaggregating control and data planes [5]. Third, the northbound interface and API is defined in 5G, where through this interface the 3GPP network services and capabilities are exposed to third-party application/services [19].

3.5 RAN Evolution in 5G

3.5.1 The Next-Generation Radio Access Network

The first technical specifications for the 5th generation of mobile technology were released by 3GPP in June 2018. The 5G system (5GS) consists of a core network 5GC and a RAN called NG-RAN (Next-Generation Radio Access Network) [20].

It is important to notice that the NG-RAN supports both the new radio access called simply NR standing for New Radio and the evolved legacy 4G radio access –LTE–. As a consequence, NG-RAN defines two types of radio base stations: the ng-eNB, an evolution of the 4G eNB that still uses LTE on the radio interface and the new gNB providing NR access.

The gNBs and ng-eNBs connect to the 5G core network via a standardized interface called NG. As the 5G core was designed from its inception to support the CUPS, the NG interface is in fact split such that the gNB will connect to a network function named Access and Mobility Management Function (AMF) for the control plane operations and to the user plane function for the forwarding of user plane data (Figure 3.16). The NG-RAN base stations are also connected between themselves by means of an interface called Xn which may be thought as an evolution of the X2 interface described in Section 3.3.4.3 and performing an equivalent role [20, 21].

It should be noted that unlike the migration from the previous to the most recent generation that required a complete deployment of new base stations and new core network elements, 3GPP defined in late 2017 the "Non-Stand-Alone NR new radio specifications for 5G," commonly referred to as 5G NSA. The 5G NSA model enables a base station supporting

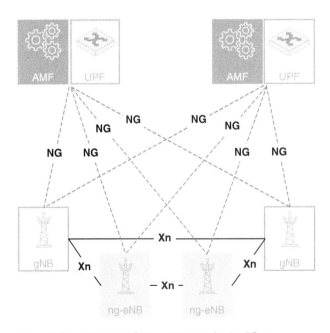

Figure 3.16 High-level RAN architecture in the 5G era.

NR to connect to the legacy EPC, that is the 4G core, and to interwork with existing 4G base stations that use LTE. 5G NSA has proven extremely popular to the point that as of the end of 2022, all operators who have deployed 5G worldwide have started the deployment of 5G RAN using this model. The reason is that 5G NSA requires small enhancements to the already existing 4G RAN and allows operators to deploy the more advanced 5G NR interface yielding higher throughput and more system capacity where it is most required.

3.5.2 gNB Split Architecture

3.5.2.1 Rationale for Split Architecture

The split of the 4G eNB into BBU and RRH still resulted in a somewhat monolithic design of the base station, and while the advantage of having multiple RRHs served by a centralized BBU was significant, the lack of tight specifications of the connection between these two elements resulted in practice in the absence of multi-vendor RAN. As discussed in Section 3.3.4.1 vendors adopted CPRI as the interface of choice between RRHs and BBU but also made use of proprietary adaptations to achieve the desired performance or to overcome ambiguity in the CPRI specifications.

During the design of the gNB, the concept of disaggregating the components was taken a step further resulting in the definition of three distinct modules: the Central Unit (CU), the Distributed Unit (DU), and the Radio Unit (RU). These modules implement all the RAN functionality that is needed to either prepare the data to be transmitted to the UE (downlink direction) or to process the incoming radio signals received from the UE and extract the user data (uplink direction). At high level, the RRH role is performed by the RU, while the BBU role is performed by the CU and DU.

More specifically, the CU takes care of the non-real-time processing of data, that is, with the data arriving from the core network (downlink) or in the last stages of the processing (uplink). As the data at this phase is not very delay-sensitive and throughput is not too extreme, a single CU may serve multiple DUs. The CU lends itself for being virtualized increasing further the scalability. Furthermore, 3GPP extended the CUPS work from the core (see Section 3.3.4.4) to the CU which consists of a user plane part (CU-UP) and a control plane part (CU-CP) connected via a standardized interface (E1). 3GPP also defines a standardized interface between the CU and the DU called F1 which is also separated into a user plane (F1-U) and control plane (F1-C) component (Figure 3.17).

The Distributed Unit and the Remote Unit implement functionality that requires very low latency as well as the processing of very high bandwidth signals. It should be noted that, unlike the CU-DU split, 3GPP provides possible recommendations on which functionality of the RAN should be included in the DU and which functionality should be included in the RU. Generally speaking, however, the goal is for one DU to be able to serve multiple RUs as well as to be able to virtualize most, if not all the DU functionality so that it can be deployed in data centers over COTS hardware. The trade-off on where to split the functionality between RU and DU depends on many factors, and a large body of studies exists on this subject. Performing more signal processing in the RU decreases the bandwidth required for the fronthaul, that is, the connection between RU and DU which could otherwise easily jump to tens of Gbps in a fully developed 5G system. Besides the fronthaul bandwidth, another important consideration for the deployment of a 5G split architecture

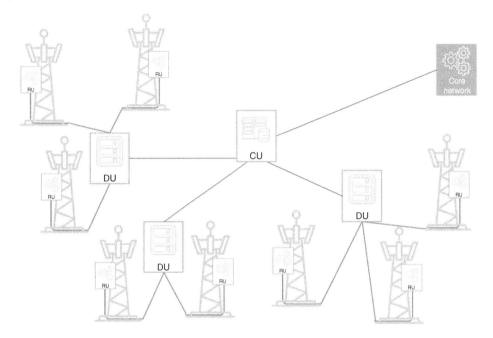

Figure 3.17 Representation of RAN using gNB split architecture.

is that since the RU cannot be virtualized, minimizing the cost and energy consumption of this component will result in an overall reduction of the total cost of ownership of a 5G RAN. In Chapter 4, we will study in more detail how O-RAN Alliance has tackled the split of functionality between CU, DU, and RU.

In summary, the 5G split architecture and virtualization of parts of the RAN not only yield a reduction in the overall cost of deploying the RAN but also serve the purpose of making 5G more adaptable and responsive to market needs. The introduction of LTE Advanced (LTE-A) and even more so of 5G has seen the proliferation of different use cases for the mobile system each coming with its own set of requirements. Whereas the consumer market is for the time-being still dominated by three highly standardized services (voice, messaging, and data), a wide variety of new use cases is expected to emerge as 5G is further deployed, each use case may have its own set of bespoke connectivity needs and KPIs that may not be aligned with those of consumer services. For example, while to serve a mobile broadband user, the most important parameter is currently download speed, most basic IoT devices require the RAN to deliver instead a stable uplink over large areas and lowest possible power consumption; at the same time, devices operating in a Multi-Access Edge Computing environment may prioritize very low latency or high reliability over speed of the connection.

As already observed in 4G, a RAN design and deployment that is expected to be accessed by different populations of users is often the result of attempting to balance the requirements of the typical consumer of mobile data communication services and the requirements of IoT. As with any compromise, the result is a system that is not optimally using all its capabilities. Through disaggregation of the RAN functionality and by moving these functions from hardware to software, it becomes possible for the 5G access network to be configured in such a way as to be optimal for the intended usage. This is the fundamental idea behind the concept of Network Slicing which started at the core network level and which is slowly but surely migrating to also include the 5G RAN.

Another advantage of a virtualized RAN (vRAN) is to be able to allocate resources where they are needed. In Section 3.2.2, we mentioned that the RAN is not dimensioned for the peak traffic load as this would result in a too high waste of resources. With virtualization, it becomes possible to divert resources to where they are needed and therefore increase the utilization factor. For example, a CU managing cell sites in a residential and business area can follow the natural traffic pattern and reduce the number of connections that need to be dropped due to the lack of capacity: this particular pattern is often called tidal effect.

3.5.2.2 Split Architecture Topology Options

When deploying a RAN using the split RAN units, operators are presented with a spectrum of options going from a fully distributed RAN to a fully centralized topology. These are shown in Figure 3.18. The selection of the optimal topology depends on several factors such as the extent of centralization possible considering latency-sensitive functionality, the availability of transport infrastructure, the complexity, and cost among others. The RAN topology can be characterized at high level on the basis of which of the eight options defined in 3GPP to split the RAN has been implemented, in particular:

- A configuration where the CU is separated geographically from the rest of the RAN is said to be a High-Level Split (HLS). The HLS using option 2 is fully specified in 3GPP.
- A configuration where the separation of functionality is closer to the antenna, that is, at the lower OSI layers, is said to be a Low-Level Split (LLS). The LLS is the preferred approach for organizations such as O-RAN Alliance and is analyzed in more detail in Chapter 4

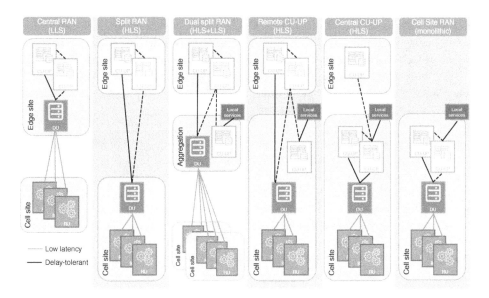

Figure 3.18 Split RAN topologies. Source: Reproduced from [22].

Table 3.1 Distribution of CU, DU, and RU for different topologies and inter-site interfaces.

Placement option	Cell site	Aggregation site	Edge site	Inter-site interfaces
Central RAN (LLS)	RU		DU, CU	LLS
Split RAN (HLS)	RU, DU		CU	F1c, F1u
Dual-split RAN	RU	DU, [CU-UP]	CU	LLS then F1c, F1u, [E1]
Remote CU-UP	RU, DU, [CU-UP]		CU-CP, CU-UP	F1c, F1u, [E1]
Central CU-UP	RU, DU, CU-CP, [CU-UP]		CU-UP	F1u, E1
Cell site RAN	RU, DU, CU			S1 and/or NG

It should be observed that more that a RAN deployment may expose both HLS and LLS as it is the case of a fully disaggregated RAN with CU, DU, and RU geographically separated from one another.

NGMN expanded on the possible topologies by providing several examples configurations [22]. With reference to Figure 3.18, Table 3.1 describes the placement of each of the split RAN units and what interface is used by a site to connect to another site.

3.5.3 Fronthaul Evolution: eCPRI

With the advancements of the 4G radio access technology from LTE to LTE Advanced and LTE Advanced Pro, it became possible to increase substantially the maximum throughput of a base station. This increase was partly due to more sophisticated digital signal processing strategies, but an important factor was the possibility of increasing the number of antennas used in antenna system as well as the re-purposing of frequency bands initially utilized for other mobile generations. The consequence of larger bandwidths and of having multiple antennas in each of the base station sector had a significant impact on the link between RRH and BBU. With 5G set to further increase throughput and introduce even more elaborate antenna systems with tens of elements (Multiple Input Multiple Output systems), it became clear that CPRI was no longer fit for purpose [23].

In response to this challenge, in 2017, a group of infrastructure vendors including Ericsson AB, Huawei Technologies Co. Ltd, NEC Corporation, and Nokia released the specifications of the evolution of the CPRI protocol called eCPRI [24].

The new protocol based on Ethernet adopts more advanced synchronization techniques but more importantly allows a 10-fold reduction of the bandwidth requirements, reduces latency, and jitter thanks to the use of packet-based transport, specifically Ethernet or IP.

Besides defining evolved versions of the REC and RE, respectively, called eREC and eRE, eCPRI also introduces networking elements that allow more flexibility in the deployment of the connection between the RRH and the BBU. For example, unlike its predecessor, eCPRI supports multi-point to multi-point connections where several eRECs can connect to several eREs.

The other main innovation of eCPRI compared to CPRI is that it is possible to realize different split of the physical layer functionality between the eREC, located at the BBU and the eRE located close to the antenna system. eCPRI defines three possible splits of the physical layer functions with each split resulting in different demands on the throughput and synchronization features of the transport network between eREC and eRE. Depending on the specific situation, it is possible to move most of the physical layer functionality at the eRE allowing for less stringent requirements on the transport network and higher distance between RRH and BBU or conversely, moving most of the physical layer functions in the eREC simplifying the design and costs of the eRE at the expense of the fronthaul complexity. Concentrating physical layer functionality in the eREC is generally preferred since it allows to add new base station functionality by upgrading the BBU which is logistically simpler and more effective than replacing the RRHs considering that a single BBU can serve several antenna systems. Some of the features that the physical layer functionality split allows include carrier aggregation (CA), downlink coordinated multi-point transmission and reception (CoMP), and management of MIMO (multiple input multiple output) (Figure 3.19).

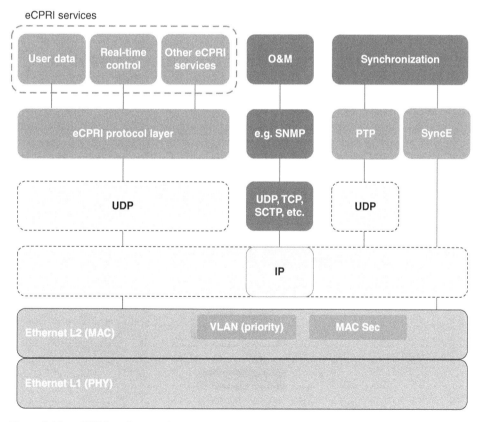

Figure 3.19 eCPRI interface architecture. Source: Reproduced from [24].

3.6 Evolution of the Base Station Architecture

3.6.1 Overview

The base station architecture has considerably evolved, especially in the past few years, and the traditional vendors are for the most heading toward providing a RAN which leverages both standards innovations and the IT paradigms described in Section 3.4. This architecture is commonly referred to as vRAN. We will trace in this section the different milestones in the evolution toward vRAN before describing some of the salient features of vRAN. This is not simply to provide a historical context, but rather because, depending on network topology, geography, desired RAN KPIs and other factors, the overall RAN deployed by an operator may use several of the configurations described. The high-level visualization of the evolution of the RAN architecture is depicted in Figure 3.20 where it can be noted that the difference is in how the RRH and the BBU are split. The reader can refer to Section 3.3.4.1 for a detailed description of this split.

3.6.2 Distributed RAN

In the distributed RAN, the RRH and BBU are co-located. The distributed RAN is the more traditional type of base station typically deployed to provide wide area coverage with the antenna system mounted on a mast. The advantage of this configuration is that the relatively small distance between the RRH and the BBU does not pose major challenges for the fronthaul and simplifies the coordination of the RRH and BBU operations (Figure 3.21). On the other hand, the disadvantage is that each cell site needs to be equipped with both units driving the cost of the cell site up [25].

This configuration was the only one available to be used in the early days of mobile communications, with 2G base stations usually hosting both RRH and BBU on the same site. This made sense in the 2G era because the spectrum used was in the lower bands (usually lower than 1GHz) and the major use cases served were voice calling and short messaging

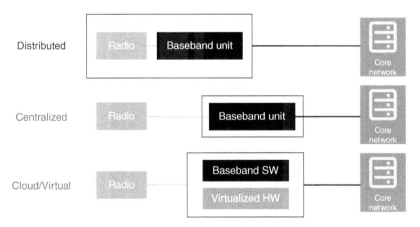

Figure 3.20 Simplified visualization of the evolution of RAN.

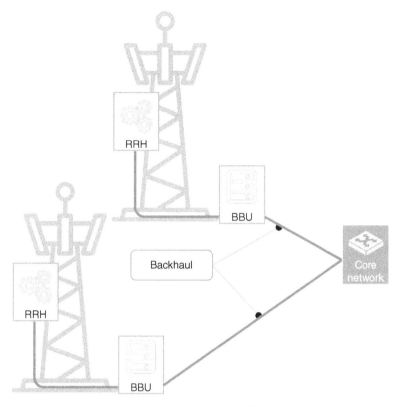

Figure 3.21   System architecture with distributed RAN.

services. Therefore, it was possible to establish cell sites that can cover a wide range with the lower spectrum bands while satisfying the requirements of low bandwidth services of voice calling and messaging.

However, it has proven increasingly difficult to acquire new cell sites for the deployment of a distributed RAN base station due to the size of the footprint and costs of the passive elements. It should also be noted that an operator using a distributed RAN architecture for each of the generations in use (e.g. 2G, 3G, and 4G) will need theoretically to deploy separate RRH and BBUs for each of the systems.

3.6.3   Centralized RAN

In the centralized RAN, the RRH and BBU are not co-located. Instead, the RRHs are deployed remotely next to the antenna system and are connected via a fronthaul interface to a BBU that processes the radio signals from several RRHs belonging to one or several base stations. In this configuration, the radio cell sites are *small*, as they only have RF-related functionalities, making them viable to be placed in more compact and niche places (e.g. indoor settings) something that would have been impossible or too costly to establish a distributed RAN base station. The disadvantage of this configuration is that it

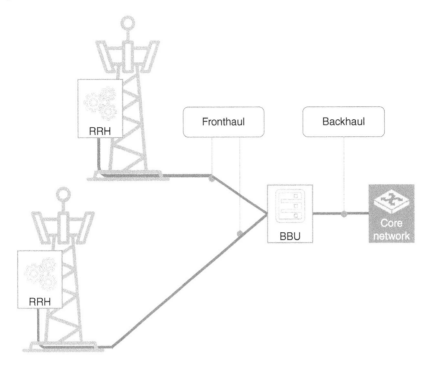

Figure 3.22 System architecture with centralized RAN.

is relatively complex to coordinate the separated RRH and BBU, and that the connection from the RRH and BBU, that from the few meters of a distributed RAN base station may now be several kilometers requires relatively high bandwidth, often only available with a fiber connection (Figure 3.22).

This configuration gained traction as the mobile communication generations addressed the data use cases, especially in the 4G era. The data use cases, unlike the voice calling use cases, require high bandwidth which requires higher spectrum bands. This means that the coverage of the cells was reduced, and the operators were required to establish distributed RAN base stations more densely. However, densification of the network may be physically impossible under some circumstances (e.g. indoor settings) or economically infeasible due to cost considerations. Therefore, the centralized RAN configuration was adopted, allowing several RRHs to be installed with high density and providing enough coverage and capacity.

In a centralized RAN architecture, the norm is that the RRH and BBU belong to the same vendor.

Cloud RAN is a variation of centralized RAN, where the BBU is virtualized and is provided over cloud infrastructure. This configuration helps to leverage cloud infrastructure with economies of scale and to benefit from more flexible deployment of network functions (Figure 3.23). It is also sometimes referred to as virtualized RAN, which is elaborated further in the subsequent section.

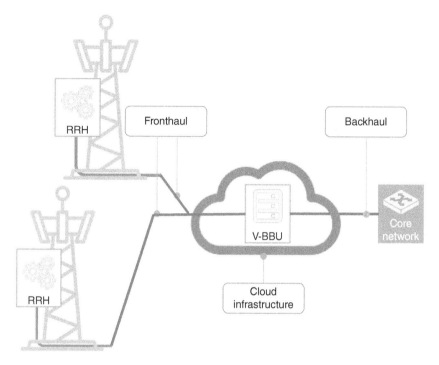

Figure 3.23 System architecture with Cloud RAN.

3.6.4 Virtualized RAN (vRAN)

3.6.4.1 Concept
In the vRAN, the split of the RU and BBU is the same as the centralized RAN. However, the key difference is that principles of virtualization, cloudification, and SDN are applied to the BBU. Therefore, the BBU's hardware and software are decoupled to yield the following benefits:

- Software upgrades without having to replace hardware.
- Adoption of mature IT technology paradigms such as DevOps, SDN, and life cycle management.
- Flexible scaling of the control plane and the user plane, without having to scale one because of the expansion of another.
- Efficient utilization through pooling (example shown in Figure 3.24).

One important difference when the IT industry techniques are applied in the telecommunications industry is that the telecommunications industry tends to have much more stringent requirements in terms of availability and reliability. For example, it is common for an operator to require telco-grade reliability such as five-nines (i.e. which translates in the system being unavailable for a maximum of 27 seconds in a year) and resiliency. Due to the expectation that mobile operators meet such requirements and that historical data suggest that IT systems are not meeting the telco-grade criteria, a pure form of vRAN where everything is virtualized is still presenting some challenges. Furthermore, while the

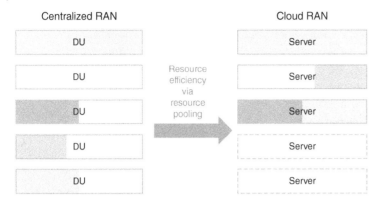

Figure 3.24 Resource utilization with pooling.

ultimate goal is the complete softwarization of the base station digital signal processing, some of the most computationally complex algorithms such as channel estimation, power control, scheduling, and fast Fourier transform are still more efficiently realized using dedicated hardware accelerator rather than software running on the commercial off-the-shelf hardware which is needed to access the several advantages of virtualization described earlier.

References

1 GSMA, "Legacy mobile network rationalisation: experiences of 2G and 3G migrations in Asia-Pacific," 2020.

2 3GPP, "3GPP TS 36.401 v17.0.0 LTE; E-UTRAN; Architecture Description".

3 CPRI, "Common Public Radio Interface (CPRI); Interface Specification", 2013.

4 KeySight, "O-RAN Next-Generation Fronthaul Conformance Testing (whitepaper)," KeySight, USA, 22 July 2020.

5 P. Schmitt, B. Landais and F. Y. Yang, "Control and User Plane Separation of EPC nodes (CUPS)," 3 July 2017. [Online]. Available: https://www.3gpp.org/cups. [Accessed 9 May 2022].

6 O-RAN Alliance, "O-RAN: Towards an Open and Smart RAN," O-RAN Alliance, October 2018.

7 ETSI, "ETSI White Paper No. 23: Cloud RAN and MEC: A Perfect Pairing," ETSI, February 2018. [Online]. Available: https://www.etsi.org/images/files/ETSIWhitePapers/etsi_wp23_MEC_and_CRAN_ed1_FINAL.pdf. [Accessed 30 August 2020].

8 NIST, "Virtualization - Glossary CSRC," [Online]. Available: https://csrc.nist.gov/glossary/term/Virtualization. [Accessed 13 April 2022].

9 ETSI NFV ISG, *GR NFV 003 v1.6.1,* Sophia Antipolis: ETSI, 2021.

10 NTT DOCOMO, *"5G Open RAN Ecosystem Whitepaper,"* 2021.

11 Techplayon, "Open RAN (O-RAN) Reference Architecture," Techplayon, 28 October 2018. [Online]. Available: http://www.techplayon.com/open-ran-o-ran-reference-architecture/. [Accessed 10 August 2020].

12 ETSI NFV ISG, "Network Function Virtualization - Introductory White Paper," in *SDN and OpenFlow World Congress*, Darmstadt, 2012.

13 ETSI NFV ISG, "GS NFV 002 v1.2.1," ETSI, 2014.

14 NIST, "The NIST Definition of Cloud," 2011.

15 A. Reznik, "What is Edge?," 2018. [Online]. Available: https://www.etsi.org/newsroom/blogs/entry/what-is-edge. [Accessed 5 May 2022].

16 Y. C. Hu, M. Patel, D. Sabella, N. Sprecher and V. Young, "Mobile Edge Computing A key technology towards 5G," ETSI, 2015.

17 Open Networking Foundation, "Software-Defined Networking (SDN) Definition," [Online]. Available: https://opennetworking.org/sdn-definition/. [Accessed 9 May 2022].

18 K. Benzekki, A. E. Fergougui and A. E. Elalaoui, "Software-defined networking (SDN): a survey," *Security and Communication Networks,* no. 9, pp. 5803-5833, 2016.

19 Y. Yan, *New WID on Rel-17 Enhancements of 3GPP Northbound Interfaces and Application Layer APIs,* 3GPP, 2021.

20 3GPP, "3GPP TS 38.300 v17.0.0 5G; NR; NR and NG-RAN Overall Description," 2022.

21 3GPP, "3GPP TS 38.401 v17.0.0 5G; NG-RAN; Architecture Description," 2022.

22 NGMN, "NGMN Overview on 5G RAN Functional Decomposition," NGMN, 24 February 2018. [Online]. Available: https://www.ngmn.org/wp-content/uploads/Publications/2018/180226_NGMN_RANFSX_D1_V20_Final.pdf. [Accessed 7 October 2023].

23 ITU-T, "GSTR-TN5G: Transport network support of IMT-2020/5G," ITU, February 2018. [Online]. Available: https://www.itu.int/dms_pub/itu-t/opb/tut/T-TUT-HOME-2018-PDF-E.pdf. [Accessed 30 August 2020].

24 Ericsson AB, Huawei Technologies Co. Ltd, NEC Corporation and Nokia, "eCPRI Specification V2.0: Common Public Radio Interface," 10 May 2019. [Online]. Available: http://www.cpri.info/downloads/eCPRI_v_2.0_2019_05_10c.pdf. [Accessed 7 October 2023].

25 K. Murphy, "Centralized RAN and Fronthaul," Ericsson, [Online]. Available: https://www.isemag.com/wp-content/uploads/2016/01/C-RAN_and_Fronthaul_White_Paper.pdf. [Accessed 6 July 2019].

26 M. Shelton, "Open RAN Terminology," Parallel Wireless, 20 April 2020. [Online]. Available: https://www.parallelwireless.com/open-ran-terminology-understanding-the-difference-between-open-ran-openran-oran-and-more/. [Accessed 29 July 2020].

4

O-RAN Alliance Architecture

4.1 High-Level Objectives of the O-RAN Alliance Architecture

As presented in Chapter 2 the O-RAN Alliance has emerged as possibly the most influential organization in the area of Open RAN. O-RAN Alliance was formed in 2018 to promote the commercialization of a more open radio access network (RAN) and accelerate innovation. At high level, the architecture of Open RAN aims to fulfill the following goals:

- Foster the creation of a multi-vendor environment.
- Introduce in the RAN modern IT paradigms.
- Apply programmability to the RAN.
- Leverage the rise of artificial intelligence and machine learning to improve the RAN performance.

By focusing on the openness of the interfaces and components, O-RAN Alliance aims to create a diverse ecosystem where the different parts of the radio access network, both hardware and software, can be procured from different vendors, can be integrated easily, and can achieve the same or better performance compared to single-vendor implementation.

As well as achieving parity in performance an important design principle for the O-RAN Alliance has been to preserve full compatibility with the overall 3GPP system, in particular this means that the O-RAN Alliance RAN will expose interfaces toward the core network and toward the UE that are fully compliant to 3GPP specifications. This will enable operators to deploy the O-RAN Alliance RAN without the need to make modifications to rest of the system and especially to maintain full compatibility with UEs.

4.2 O-RAN Alliance Work on 4G

4.2.1 Laying the Ground for a Multi-vendor Environment

While O-RAN Alliance is primarily focused on 5G, it also created specifications and profiles for the 4G system. This decision was driven by the consideration that most of the early deployments of 5G are based on the so-called "non-standalone," meaning that early 5G systems heavily relied on the 4G core network and tight interworking with the 4G radio access network. To better understand the rationale for the profiling of some of the 4G interfaces

Open RAN Explained: The New Era of Radio Networks, First Edition.
Jyrki T. J. Penttinen, Michele Zarri, and Dongwook Kim.
© 2024 John Wiley & Sons Ltd. Published 2024 by John Wiley & Sons Ltd.

and why it was deemed essential to achieve the objective of creating a multi-vendor ecosystem, it pays to take a closer look at the possible deployment options of a 5G.

4.2.2 Standalone and Non-standalone 5G

Traditionally a mobile system generation is characterized by a new RAN technology aiming at meeting a set of performance requirements issued by the International Telecommunications Union (ITU), and a core network. In the case of the 4G mobile system, the newly defined RAN is called Evolved Universal Terrestrial Radio Access Network (E-UTRAN), while the newly defined core network called is called Evolved Packet Core (EPC). Operators needed to deploy both E-UTRAN and EPC to provide 4G services and to be able to meet the ITU IMT Advanced requirements [1].

The 4G mobile system can also interwork with previous mobile generations, but this interworking takes place through a connection of the 4G and legacy systems at core network level, and it is not different from the interworking with any other data or telephony network. In other words, the 4G RAN has no direct interaction with any other system. This is the traditional way a mobile network is deployed, and it is commonly referred to as standalone deployment.

At the time when 5G was standardized, 3GPP departed from the principle of generations working in isolation and interworking at core network level, defining instead several deployment models that were combining elements of the 4G system and 5G system.

The deployment models envisaged by 3GPP are classified as standalone (SA) and non-standalone (NSA) depending on whether single or two generation(s) of radio access technology are employed.

Figure 4.1 shows five possible configurations of the mobile system that uses either the newly defined 5G Core (5GC) or the newly defined 5G radio access node (NR). Out of these five models, 2 and 5 are standalone deployments (that is only one radio access technology is used), while models 3, 4, and 7 are NSA since the RAN consists of both 4G base stations (using LTE radio access technology) and 5G base stations (using NR radio access technology). For completeness, Figure 4.1 also illustrates the standalone deployment where the 4G core network connects to the RAN consisting of only LTE base stations: this is the "regular" 4G mobile system deployed around the world.

Of the standalone scenarios, Option 2 is the one corresponding to the traditional deployment strategy for the past generations: the newly defined 5GC is connected to the newly defined 5GNR RAN. This option, often simply referred to as 5G-SA, is the target configuration for many of the major mobile operators around the world and is being gradually being rolled out in by many operators worldwide. The other standalone deployment depicted in Figure 4.1 is Option 5. This option would allow an operator to control the already well-established E-UTRAN using the new generation 5GC accessing several of the benefits that the 5GC introduces with relatively inexpensive radio network upgrades.

Of the three NSA configurations, the most important is Option 3 which is in fact commonly referred to just as 5GNSA. As all the NSA configurations, Option 3 uses both 4G and 5G radio access technologies with 4G radio access being the primary access and 5G secondary access. In this deployment model, both 4G and 5G base stations are controlled

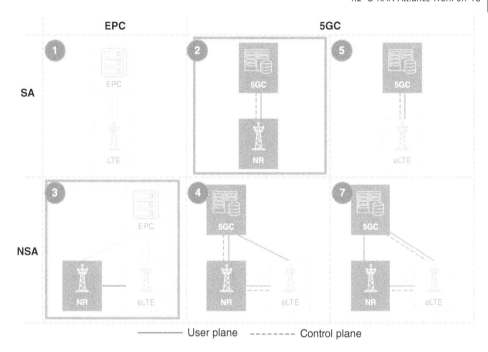

Figure 4.1 Deployment models for 5G. Source: Data taken from GSMA, 5G Implementation Guidelines [2].

by EPC which only requires minor upgrades since the decision on whether to use 5G or 4G radio access to transfer user data is made in the RAN. The rationale behind 5GNSA is that existing 4G deployment (4G RAN and the 4G core) can be leveraged for providing existing service, while 5G NR base stations supplement the 4G coverage and provide a possible boost in speed and capacity where most needed (e.g. busy urban areas). Operators can therefore gradually introduce coverage of 5GNR where it is most needed, upgrading where possible existing sites without risking of negatively impact the service availability for users of the 4G system. Besides, whereas in the 5G SA, Option 2 deployment model the UE needs to be upgraded to support both the NR front end and the 5GC new protocols, the UE accessing a NR base station deployed according to 5GNSA Option 3 only requires to support the new radio access protocol.

The other NSA options received initially some support, but 3GPP decided not to complete the full specifications for them and are therefore unlikely to be used in commercial deployments.

4.2.3 The O-eNB and 4G Interfaces Profiling

4.2.3.1 O-eNB

In the most common early 5G deployment, the 5GNSA discussed in Section 4.2.2, the RAN consists of both upgraded 4G LTE nodes (eNodeBs) and 5G NR nodes (gNodeBs). The requirement on the upgraded eNodeBs to be deployed as part of a 5GNSA setup is to be able to support 5G interfaces and protocols on top of the regular 4G interfaces and protocols. In 3GPP, the such eNodeB enhanced to work alongside gNodeB is called ng-eNB.

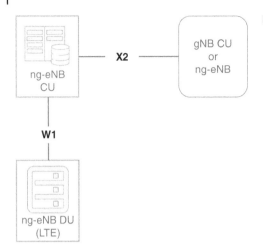

Figure 4.2   ng-eNB standardized interfaces.

Just as the gNodeB, the ng-eNB follows the split architecture principles that 3GPP defined in Release 15 and that was presented in Chapter 3. This implies that unlike the 4G eNodeBs that are defined in standards as monolithic design, the ng-eNB consists of three different modules: the Radio Unit (RU), the Distributed Unit (DU), and the Centralized Unit (CU). The ng-eNB supports the X2 interface toward another ng-eNodeB or toward a gNodeB, and a standardized interface is specified between the ng-eNB CU (Figure 4.2) and the ng-eNodeB DU called W1 [3].

As part of the work on 4G, the O-RAN Alliance introduced in its specifications the concept of O-eNB, a radio access node that can be deployed to serve as a regular eNodeB or as an ng-eNB. As for the ng-eNB, the O-eNB consists of a DU called O-DU and a CU called O-CU, which are connected using the standardized W1 interface.

The O-eNB is connected to the core network either via S1 interface when operating as a eNodeB or via the NG interface, and for this reason the O-eNB needs to support the traditional LTE protocols (such as the RRC, PDCP, RLC, MAC, and PHY) as well as additional 5G NR protocols (such as the SDAP and NR PDCP).

4.2.3.2   X2 and W1 Profiling

In the 3G architecture, groups of base stations were connected to each other via a node called RNC. However, when 4G RAN architecture was designed, it was recognized that connecting two base stations to each other directly, rather than via an intermediary node such as the RNC or, as in the case of 2G, via the core network yielded many advantages. Direct connection between two adjacent base stations is particularly beneficial in the scenarios where the UE is located in the coverage area of these two base stations or when the UE is moving from one cell of a base station to another cell of a connected base station.

3GPP defined in the 4G specifications a new interface called X2 for the purpose of allowing two eNBs to communicate directly with each other [4]. As it has been the case of other standardized interfaces in the past, RAN vendors added a number of proprietary extensions that optimized the exchange of information between two of their eNodeBs, and as a consequence, while it is still possible to connect base stations from different vendors via X2, this results often in a performance degradation compared to the scenario where two base stations from the same vendor were connected together. The consequence is that operators who use more than one RAN vendor often deploy each vendor in a specific region and do not connect eNodeBs across the region border.

As one of the main objectives of O-RAN Alliance is to foster a multi-vendor environment, work was carried out to address this problem, resulting in the definition of a profile for the X2 interface [5]. Unlike a full specification, the purpose of a profile is to improve the usability of a standard, especially in the case where standards allow for many different options causing interoperability problems.

The profiling of X2 is also very significant in the context of the 5G NSA deployment since it signaled the desire of the members of O-RAN Alliance (mostly mobile network operators) to ensure that a 4G eNodeB and 5G gNodeB from different vendors could be successfully deployed, removing the potential lock-in with the existing 4G eNodeB vendors.

Along with X2, O-RAN Alliance also created a profile of W1, the interface that connects the eNB DU with the eNB CU [3]. As for X2, the profiling of W1 by O-RAN Alliance is meant to enable operators to choose different RAN vendors for supplying different parts of the base station; however, by the time W1 was specified, most of the operators already had deployed eNBs so this specification has been less impactful.

4.3   O-RAN 5G Architecture

4.3.1   Architecture Overview

The O-RAN architecture diagram is shown in Figure 4.3. The readers accustomed to the 3GPP Release 15 RAN architecture will note several familiar nodes and interfaces: this is not a coincidence since the baseline of the O-RAN architecture is indeed the 3GPP Release 15 "split architecture" that we introduced in Chapter 3. Specifically, it is possible to recognize in the figure the three elements of the gNodeB: CU, DU, and RU. Owing to the fact that unlike 3GPP, O-RAN Alliance has a stricter definition of what functionality belongs to each of these units and how they interface to each other, they are renamed in the O-RAN Alliance documentation as O-CU, O-DU, and O-RU. O-RAN Alliance also added new network elements and interfaces to its RAN Architecture that do not have an equivalent in 3GPP.

The elements of the O-RAN Alliance architecture and the interfaces connecting these elements are described in this section.

NOTE: the O-RAN specifications are being actively developed. In this book we refer to the Release 2 set of specifications. The latest versions of the O-RAN Alliance specifications are available on the O-RAN Alliance website at this address: https://orandownloadsweb .azurewebsites.net/specifications

Note: As the main focus of this section from now on is the work of O-RAN Alliance in the context of a 5G technology and specifically on the RAN consisting of gNodeB, Figure 4.3 does not include the O-eNB described in Section 4.2.3.1. The O-eNB, however, can also be managed by the Service Management and Orchestration (SMO) Framework and connected to the Near-Real-Time RIC via the E2 reference point and to other eNBs/gNBs via the X2/Xn interfaces.

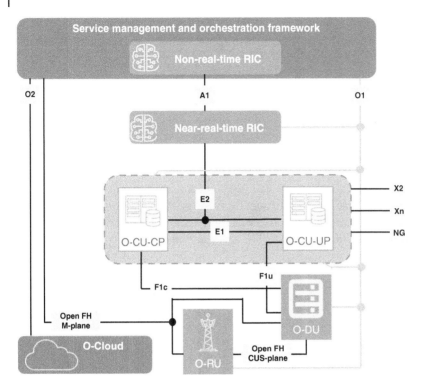

Figure 4.3 O-RAN logical architecture. Source: Adapted from O-RAN Alliance [5].

4.3.2 O-RAN Alliance Architecture Network Components

4.3.2.1 O-RAN Central Unit (O-CU)

The O-CU [5] is the CU profiled and enhanced by O-RAN Alliance. The O-CU can be further decomposed into a control plane part (O-CU-CP) and a user plane part (O-CU-UP). In the default O-RAN Alliance architecture, the O-CU implements functionality up to the Packet Data Convergence Protocol (PDCP) of the RAN protocol stack. The interface E1 between CU-CP and CU-UP defined by 3GPP is re-used by O-RAN Alliance in order to ensure compatibility.

The O-CU terminates the backhaul, and maintains the connection with the core network via the NG reference point consisting of an interface toward the Access and Mobility Function (AMF) and toward the Session Management Function (SMF).

The O-CU also connects with the O-DU using the 3GPP specified F1 interface which is split into the control plane part (F1-C) between the O-CU-CP and the O-DU and the user plane part (F1-U) between the O-CU-UP and the O-DU.

4.3.2.2 O-RAN Distributed Unit (O-DU)

The O-RAN Alliance Distributed Unit (O-DU) [5] is a profiled and enhanced DU as defined by 3GPP. Unlike 3GPP which does not strictly define the boundary between DU and RU, O-RAN Alliance recommends a specific distribution of the RAN functions between O-DU and O-RU. In this configuration, the O-DU contains the Radio Link Control (RLC), the

Medium Access Control (MAC) layer, and part of the physical layer of the RAN protocol stack. The O-DU connects to the O-CU-CP via the 3GPP-defined F1 interface and with the O-RU via the Open Fronthaul protocol. The detailed description of the functional split between O-RU and O-DU is presented in Section 4.4.3.3.

4.3.2.3 O-RAN Radio Unit (O-RU)

The O-RAN Alliance Radio Unit (O-RU) [5] is a logical node that implements the functions of the low physical layer of the RAN protocol stack together with the Radio Frequency processing. The O-RU resembles the traditional Remote Radio Head (RRH) but may also perform some additional physical layer processing, thereby reducing the bandwidth required for the fronthaul, that is, for the connection between O-RU and O-DU.

4.3.2.4 Service Management and Orchestration Framework

The **Service Management and Orchestration Framework** (SMO) [5] is a new function defined by O-RAN Alliance that is responsible for running the virtualized RAN and for the configuration and management of the RAN elements. It performs also all the functions that are performed today by the Network Management System (NMS) as part of the RAN Operations Support System (OSS).

4.3.2.5 O-Cloud

The **O-Cloud** [6] is a cloud computing platform that consists of physical infrastructure nodes that meet the requirements of O-RAN specifications to host the relevant network functions (e.g. O-CU, O-DU, and Near-RT RIC), the supporting software components (e.g. Operating System, Virtual Machine, Container, etc.), and the management and orchestration functions.

4.3.2.6 RAN Intelligent Controller

The **RAN Intelligent Controller** (RIC) [7] is a function that leverages possible artificial intelligence and machine learning (AI/ML) to optimize the performance of the RAN components. More specifically, the RIC consists of two elements: the **Non-Real-Time RAN Intelligent Controller,** which is part of the SMO, and is characterized by operations with a cycle of more than one second and **Near-Real-Time RAN Intelligent Controller,** operating at shorter timescales [7].

4.3.3 Interfaces

4.3.3.1 3GPP-Defined Interfaces

In order for O-RAN Alliance compliant products to be used as part of a 3GPP system, O-RAN Alliance architecture supports the NG interface to realize the connection to the 5GC. In addition, the internal interfaces between RAN elements such as the E1 interface between the User Plane (U-Plane) and the Control Plane (C-Plane) parts of the O-CU, the F1 interface between the O-CU and the O-DU, and the Xn interface connecting two gNodeBs are applicable to the O-RAN Alliance architecture without any divergence. The X2 and W1 interfaces are also re-used in the O-RAN Alliance architecture, though they are required to comply to the O-RAN Alliance profile as discussed in Section 4.2.3.2.

4.3.3.2 O-RAN Alliance-Defined Interfaces

As well as re-using the 3GPP interfaces, O-RAN Alliance has also defined interfaces between the new components it has introduced in its architecture. As illustrated in Figure 4.3, the following reference points support the deployment of O-RAN Alliance RAN:

- *A1 interface*: This interface connects the SMO and more specifically the Non-Real-Time RIC to the Near-Real-Time RIC and is used to convey information such as machine learning models or general measurements between the two entities [8].
- *E2 interface*: The E2 interface enables the Near-Real-Time Intelligent Controller to manage the O-CU and O-DU. This interface is also used to connect to an O-eNB [9].
- *O1 interface*: The O1 interface enables the SMO to manage all the three elements of the O-RAN Alliance gNodeB (O-CU, O-DU, and O-RU) as well as the Near-Real-Time RIC [10, 11].
- *O2 interface*: Specifically designed for the operations and management of the O-Cloud [12].
- *Open Fronthaul interface*: Based on enhanced Common Public Radio Interface (eCPRI), this is the interface that enables the transfer of data between the O-DU and the O-RU and is subdivided into C-Plane, U-Plane, and synchronization plane (S-Plane). It also includes a management plane (M-Plane) that terminates in the SMO [13].

A detailed description of the format and protocols used over these interfaces is outside the scope of this chapter, however, their use will be described where appropriate when discussing the elements terminating these interfaces.

4.4 O-RAN Alliance Architecture Innovation

4.4.1 O-RAN Alliance Two-Pronged Approach

Although to date, the O-RAN Alliance architecture remains fully compatible with 3GPP architecture and can therefore be deployed alongside mobile system components compliant to 3GPP specifications, it offers a number of incremental technology advantages. These advantages can be broadly categorized in two areas:

- *Improvements to specifications and profiling*: This category includes the work on the Open Fronthaul that enabled overcoming the limitations of eCPRI; the profiling of the 3GPP interfaces between eNodeBs (X2) and gNodeBs (Xn) which opened up the possibility of overcoming some of the limitations when connecting two NodeBs from different vendors. A further major achievement that may be considered akin profiling is the endorsement of fewer options for the RAN split architecture discussed in Section 4.4.3.3.

  The value of profiling a technical specification should not be underestimated. The way standards are created in 3GPP, especially in recent years suffer from offering many different options. While the flexibility is beneficial, especially in case where there are uncertainties on the adoption of a technology or where different deployment models may exist, the reality is that options in standards have generally a detrimental effect on multi-vendor products, which, paradoxically, is the very purpose of creating a standard in the first place.

- *Introduction of advanced IT technologies and AI*: This category includes the more systematic approach to using cloud technology to support virtualized RAN network functions as well as adoption of artificial intelligence and machine learning to enhance the management and the performance of the RAN. It also starts opening up the possibility of a more software-centric approach to manage the functionality of the RAN, e.g. through the definition of Application Programming Interfaces (APIs) that can be used to "program the RAN." For this purpose, the O-RAN architecture has designed the RIC. 3GPP has started in their Release 17 to study the use of AI/ML; however, this activity is for the time being only focused on supporting operations in the 5GC, although there are plans to also include the RAN in the scope of the future releases.

Although most of the technology advancements are already commonly used by existing RAN vendors within their products, this is done in a proprietary fashion and the advantages can only be fully leveraged in a single RAN vendor environment.

4.4.2 Open Fronthaul

For historical reasons, 3GPP has never specified the interface between the RRH and the BBU as well as the interface between DU and RU. As seen in Chapter 3, for 3G and for some 4G deployments vendors adopted the Common Public Radio Interface (CPRI) specifications while eCPRI is the de facto standard for 4G and 5G. It can be observed, however, that these interface specifications are not sufficiently tight or detailed to enable true multi-vendor operations. The O-RAN Alliance stepped in and using eCPRI as baseline created the Open Fronthaul specifications that are set to become the new industry standards. Detailed description of the O-RAN activity is provided in Section 4.8

4.4.3 Stricter Approach to RAN Functional Split

4.4.3.1 Introduction

The definition of a default recommended RAN functional split between O-DU and O-RU is one of the greatest achievements of O-RAN Alliance and a critical milestone in achieving the goal of fostering the creation of a multi-vendor environment in the RAN.

As illustrated in Figure 4.4, in the context of the work on the split architecture, 3GPP did define and study eight possible alternatives for disaggregating the base station and subdivide its functionality between CU, DU, and RU. It also went further in defining a standardized interface called F1 between the PDCP which, using the common Open System Interconnection (OSI) model, is part of the OSI Layer 3 and the RLC which is part of the OSI Layer 2 [15]. By doing so, 3GPP determined de facto that the CU will contain the PDCP and will communicate with the DU over the F1 interface.

However, 3GPP did not define precisely the functionality split between the DU and the RU. Although several different alternatives were studied in a great level of detail including providing different suboptions of Option 7 (the split between high physical and low physical layer) that do describe precisely which functionality of the physical layer that are

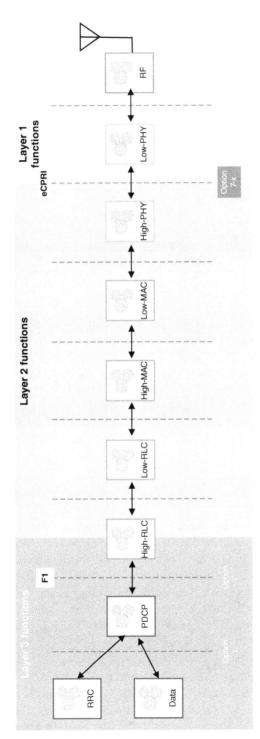

Figure 4.4 Options for splitting a base station functionality studied in 3GPP Adapted from [14].

part of the DU and those that are part of the RU, 3GPP concluded with some high-level recommendations rather than a firm guidance.

In the absence of such strict guidance, interoperability of DU and RU from different vendors would become all but impossible, since the likelihood of a DU vendor and RU vendor having selected to implement complementary functionality covering the whole RAN stack would be quite low. A number of reasons may be considered as to why 3GPP refrained from mandating a single DU-RU split, for example:

- standardization of the fronthaul interface, which would be the natural next step, was historically not performed in 3GPP with the industry relying on interfaces such as CPRI and eCPRI;
- each split of the functionality has its own advantages and disadvantages depending on the deployment scenario and other parameters so it would have been problematic for 3GPP to mandate a split that could have made some deployment scenarios difficult to realize;
- the vendors engaged in defining the 3GPP specifications did not want to restrict any option at the source and did not have the pressure to do so.

O-RAN Alliance decided to continue the work started by 3GPP, taking into consideration technical factors such as required transport bandwidth, and achievable latency, but also the complexity of the O-RU and O-DU decided to recommend a split based on 3GPP Option 7 but furnished with sufficiently detailed information to enable multi-vendor deployments. The O-RAN Alliance recommended split is referred to as Option 7-2x.

It is important to observe, however, that O-RAN Alliance recognizes that there is no such thing as one-size-fits-all and in fact acknowledges that in some scenarios (e.g. mmWave) it makes sense to include all the physical layer processing in the O-RU and split O-RU and O-DU at the MAC layer (Option 6). This specific split has been specified in detail by the Small Cell Forum which also went further defining the network functional API specification to support it.

To better appreciate the work of O-RAN Alliance in defining the functional split between DU and RU, we first need to gain a high-level understanding of the NG-RAN protocol stack.

4.4.3.2 The NG-RAN Radio Resource Management

Operations Performed in the RAN In a wireless communication system, the base station is the component responsible for transferring data to and from a mobile station using radio signals. The transfer from the base station to the mobile station is called downlink while the connection from the mobile device to the base station is called uplink.

To best utilize the radio spectrum and optimize the transmission of data to and from multiple users over a noisy channel, a base station employs a number of strategies and algorithms that are referred to with the generic term Radio Resource Management (RRM) [16]. In the 3GPP system, RRM functionality is grouped into five distinct elements that are shown in Figure 4.5. The Radio Resource Control (RRC) is also often included as part of the RRM.

Physical Layer (PHY) Roughly corresponding to the Layer 1 of the OSI model it converts digital information into radio signals and vice versa. In the uplink direction, the radio signal received at the antenna system from a mobile station undergoes a sequence of processing

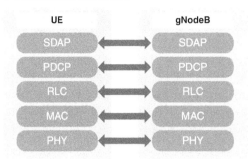

Figure 4.5 3GPP-defined base station protocol stack. Source: NR and NG-RAN\3GPP [16].

that includes conversion from analog to digital, channel estimation, and demodulation and decoding.

Likewise, in the downlink direction, the physical layer converts data intended for a mobile station into a radio signal that is structured to take into account a non-ideal radio propagation environment. The data that the physical layer generates (receives) is structured into transport channels.

Medium Access Control (MAC)   The MAC is responsible for organizing data into logical channels. In the downlink direction, for each of the mobile stations, the MAC will structure data that needs to be sent and assign a priority. It will then map the logical channels into transport channels. The MAC layer also implements error correction through Hybrid Automatic Repeat Request (HARQ), a mechanism that allows the receiver to ask for retransmission of data which has been detected as corrupted during the transmission over the air. In the uplink direction, the MAC will apply HARQ, multiplex data received from the transport channels, and create logical channels to be used by the upper layers.

Radio Link Control (RLC)   The RLC is a sublayer of the OSI Layer 2 and performs further error correction, segmentation, and desegmentation of Service Data Units.

Packet Data Control Plane (PDCP)   The main services and function of the PDCP include header compression (downlink) and decompression (uplink), ciphering/deciphering, integrity protection and verification, reordering and in-order delivery. In the OSI model, PDCP is the part of Layer 2.

Service Data Adaptation Protocol (SDAP)   This Layer 2 function performs the Mapping between a QoS flow and a data radio bearer as well as the marking of QoS flow ID (QFI) in both downlink and uplink packets.

Radio Resource Control (RRC)   RRC is the last sublayer of the RRM in 3GPP. The main services and functions of the RRC sublayer include:

- Broadcast of System Information related to AS (Access Stratum) and NAS (Non-Access Stratum);
- Paging initiated by 5GC or NG-RAN;
- Security functions, including key management;
- Mobility functions including handover and context transfer and cell reselection;

- QoS management functions;
- Management of RRC connections.

4.4.3.3 Functional Split of the Physical Layer Between O-DU and O-RU

Considering all the operations that a signal arriving at the antenna undergoes before the user data is extracted and conversely the operations to prepare data to be transmitted over the air interface to the User Equipment, it is easy to show that the most demanding part from a computational point is the functions performed in the Physical Layer.

A block diagram representation of the many operations performed in the physical layer is shown in the diagram of Figure 4.6. It is not the purpose of this chapter to discuss in detail how the data is processed at each step of its journey through the physical layer but rather to understand the rationale of O-RAN Alliance to split the physical layer functionality between the O-RU and O-DU and how this split impacts the interface between these two components of the base station. We will also compare the recommended O-RAN Alliance split to other possible distribution of the functionality between RU and DU documented in 3GPP or proposed by other organizations.

Two contrasting considerations need to be taken into account when deciding how to distribute the functionality of the physical layer between these two elements especially paying attention on how this split impacts the complexity of each unit and the required bandwidth of the interface between them (the fronthaul).

Since the RU is to be co-located with the antenna system of the base station, there are clear advantages in making the RU as simple as possible since this results in cheaper, less power hungry unit. On the other hand, the RU needs to ensure that the sheer amount of data produced by a 5G antenna system undergoes some initial processing that reduces its bandwidth to a level that can be supported by the fronthaul interface which would otherwise risk to become prohibitively expensive as well as technically challenging to the point of negating the positive effects of the disaggregation.

The compromise that O-RAN Alliance agreed upon is called Option 7-2x which is shown in Figure 4.6. This is a specific variant of 3GPP Option 7 where the split between the low and high physical layer is such that all the processing up to the fast Fourier transform (FFT) and digital beamforming is performed in the RU with the remaining physical layer processing residing instead in the DU. More specifically Option 7-2x is an optimization of the 3GPP Option 7-2 which separates the low and high physical layer at the precoding.

The difference between 3GPP Option 7-2 and O-RAN Alliance Option 7-2x is that with this split precoding is allowed to be performed either in the RU or in the DU, depending on the specific deployment scenario. There exist, therefore, two O-RAN Alliance O-RU types: the category A O-RU, which does not include precoding, and category B O-RU which is instead capable of performing precoding. While the location of the precoding has negligible impact on the fronthaul bandwidth which remains very similar to the bandwidth required with 3GPP Option 7-2, there is slight increment in the case of a category B O-RU, since the channel information that is used to configure the precoding that are worked out in the O-DU needs to be sent to the O-RU. It should also be noted that any O-DU will have to implement the precoding due to the fact that it may need to interconnect with either a category A or a category B O-RU; therefore, it is necessary for the O-DU to know when this operation should be bypassed and delegated to a category B O-RU.

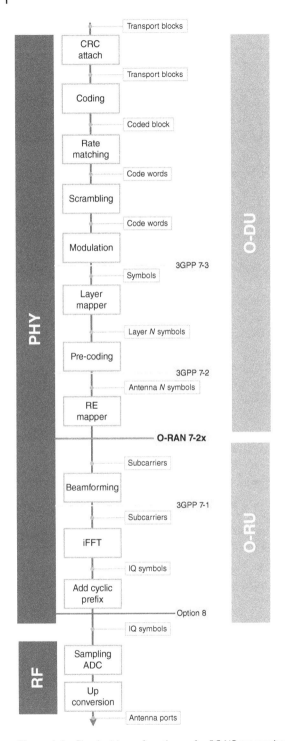

Figure 4.6 Physical layer functions of a 5G NR transmitter and receiver Adapted from [17].

The introduction of a well-defined split is instrumental for the disaggregation of the base station. The adoption of the 7-2x split enables manufacturers to focus on only developing an O-RU or an O-DU if so desired, with the confidence that it will successfully interoperate with other O-RAN Alliance compliant components.

In addition to enabling the development of a multi-vendor base station, there are also a number of technical advantages that led O-RAN Alliance to select the 7-2x split; these advantages are as follows:

- 7-2x interface scales based on "streams," which allows using high number of antennas without higher transport bandwidth. As 5G is expected to make large use of Multiple Input Multiple Output (MIMO), especially in the C-band, being able to decouple the required fronthaul bandwidth from the size of the MIMO system is a major benefit.
- With option 7-2x, there is awareness on whether the received signal contains user data or not; therefore, it is possible to avoid using transport bandwidth to carry signal that does not need to be processed.
- *Advanced receivers and inter-cell coordination*: this option allows implementation of advanced receivers and coordination features, which are also easier to implement and less restricted when most functions are placed at the O-DU. For example, UL CoMP (Coordination of Multi-point) is not possible when the UL upper-PHY processing is in the O-RU.
- *Lower O-RU complexity*: less functions at O-RU (when compared to higher split options) allow limiting the number of required real-time calculations as well as required memory requirement, especially for Category A O-RUs. This is important especially in situations where operators desire to densify the network and require the equipment at the cell site to be less expensive.
- A low complexity O-RU might even be shared between different operators favoring the neutral hosting scenarios where the infrastructure at the site premises is provided by a third party and used by multiple operators.
- *Future proof-ness*: placing most functions at O-DU will allow introduction of new features via software upgrades without inflicting HW changes at O-RU (e.g. specification changes due to URLLC or new modulation schemes).
- *Interface and functions symmetry*: if the same interface and split point is used for DL and UL, specification effort is reduced.

It is also worthwhile observing that besides the considerations about costs and complexity of the two units, the distribution of functionality between O-RU and O-DU is also determined by the technical feasibility and costs of deploying a fronthaul interface between these two nodes. Capacity of the fronthaul will have direct impact on the maximum number of O-RUs that can be connected to a single O-DU as well as the maximum distance between O-RU and O-DU.

Taking into consideration that the bandwidth of 5G signals can be quite high (100 MHz channels), that it is expected that 5G will make extensive use of MIMO antenna system, and the density of deployment requirements for indoor and urban environments, all the proposed variants of splits between DU and RU at the physical layer (that is, variants of 3GPP Option 7) place the Fast Fourier Transform operation in the RU as this is the algorithm that has the highest impact on the bandwidth. Under this assumption, it can be seen that all the 3GPP Option 7 splits as well as of course the O-RAN Alliance 7-2x split are such that the bandwidth requirements on the fronthaul can be met by using the eCPRI specifications.

4.5   Service Management and Orchestration Framework

The SMO is the entity specialized in the management of the RAN domain and it is the logical evolution of the RAN NMS. In addition, the SMO can also interface with other management domains such as the Core Network management and the Transport network management. The SMO also includes the Non-Real-Time RIC which is described in the next section.

The SMO exposes the O1 interface, the O2 interfaces as well as the Fronthaul M-Plane interface.

The O1 interface is used by the management entities within the SMO to perform configuration, operation, and maintenance and is specifically optimized for multi-vendor environment. The O1 interface is also used for FCAPS (fault, configuration, accounting, performance management, and security management).

The O2 interface is used by the SMO to communicate with the O-Cloud and enable the SMO to deploy and manage cloud infrastructure and perform lifecycle management of the workloads running in the O-Cloud.

The Open Fronthaul M-Plane interface connects the SMO to the O-RU and is used for the management of the Open Fronthaul.

A block diagram level view of the SMO is shown in Figure 4.7.

4.6   O-Cloud

The O-Cloud Platform is a set of hardware and software components which offers its capabilities to host and execute RAN virtualized functionality. Similar to other cloud

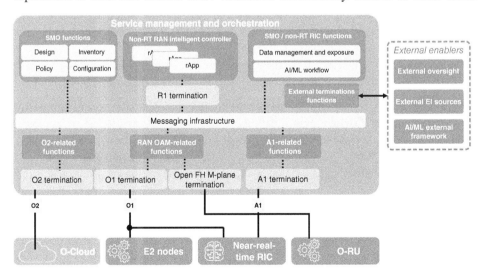

Figure 4.7   SMO framework architecture view Adapted from [18].

platforms, the O-RAN O-Cloud consists of three layers: a physical infrastructure layer, a software component, and a management and orchestration layer.

The software layer, which includes operating system and container runtime, exposes APIs enabling the management of the full life cycle of the workloads and is decoupled from the underlying hardware. The hardware layer provides computing, networking, and storage. The O-Cloud makes use as much as possible of COTS servers; however, to cater for some of the very demanding RAN functions in the physical layer, the hardware layer also comprises specialized accelerators such as ASICs, GPUs, and FPGAs. The decoupling of hardware and software allows operators to choose different suppliers, and the O-RAN Alliance specifications clearly require that the O-Cloud stack does not have any dependency on vendor-specific hardware.

Furthermore, for a cloud computing platform to comply with O-RAN Alliance, it is required that:

- It supports the O-RAN Alliance O2 interface and O-Cloud Notification interface for cloud and workload management/notifications.
- It supports the O-RAN Accelerator Abstraction Layer (AAL) API. The API enables a virtualized RAN function to discover, select, and use a hardware accelerator in the O-Cloud platform without requiring additional knowledge of vendor providing the hardware.
- It satisfies at least one deployment scenario and the associated requirements defined by the O-RAN in its Cloud Architecture and Deployment Scenarios specification.

In an O-RAN Alliance RAN architecture deployment, the O-Cloud is formed by one or more O-Cloud instances that are geographically distributed and that may support different RAN functionality. Figure 4.8 shows the components of an O-Cloud Instance which comprises:

- *One or more O-Cloud resource pools*: The O-Cloud Resource Pool is a collection of O-Clouds nodes with homogeneous profiles in one location which can be used for either Management services or Deployment plane functions.
- O-Cloud Instance Deployment Management Services (DMS) layer and Infrastructure Management Services (IMS) layer are optional components.
- *One or more O-Cloud node*: An O-Cloud Node is the abstract name for hardware resources available in the O-Cloud Resource Pool. An O-Cloud Node will consist of CPUs, memory, storage, networking, accelerators, and so on. An O-Cloud Node may offer one or more of the functionalities to the Deployment plane.
- O-Cloud Instance Deployment plane is a logical entity representing the O-Cloud Nodes across the O-Cloud Instance resources pool that is used to run a RAN network function.
- An **O-Cloud NF Deployment** is a deployment of a cloud native network function (all or partial), resources shared within a NF function, or resource shared across network functions. The NF Deployment configures and assembles U-Plane resources required for the cloud native construct used to establish the NF Deployment and manage its life cycle from creation to deletion.
- The **O2 Interface** is a collection of services and their associated interfaces that are provided by the O-Cloud platform to the SMO. The services are categorized into two

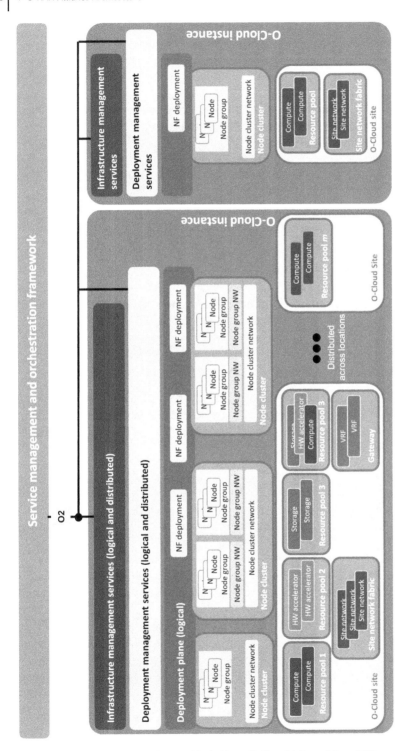

Figure 4.8 Key components involved in/with an O-Cloud Adapted from [19].

Cloud / HW features	Near-RT RIC	O-CU-CP	O-CU-UP	O-DU	O-RU
Standard cloud infrastructure (CI) and general purpose CPU	✓	✓			
CI + high-speed UP support (acceleration is optional)			✓		
CI + high-speed UP support (acceleration for O-DU)				✓	
CI + high-speed UP support (acceleration for O-RU)					✓

Figure 4.9 Cloud hardware features per component Adapted from [18].

logical groups: (i) **IMS**, which include the subset of O2 functions that are responsible for deploying and managing cloud infrastructure, and (ii) **DMS**, which include the subset of O2 functions that are responsible for managing the life cycle of virtualized/containerized deployments on the cloud infrastructure. The O2 services and their associated interfaces shall be specified in the upcoming O2 specification. Any definitions of SMO functional elements needed to consume these services shall be described in OAM architecture.

The allocation of NF deployment to a resource pool is determined by the SMO. A standard cloud infrastructure is sufficient to support the virtualized functions of the Near-Real-Time RIC, the C-Plane functions of the CU while, depending on the data rate, the virtualized O-CU-UP may require some specialized support. Moving into the lower layers of the RAN stack, due to the need of processing large amount of data with reduced delay tolerance, the use of accelerators becomes essential. As a consequence, the O-RAN Alliance O-Cloud will consist of several instances that are dimensioned and resourced appropriately for the workloads they are supporting. Figure 4.9 shows the hardware features required for the cloud infrastructure depending on the RAN node to be served.

Current platforms such as OpenStack and Kubernes deployed on a set of COTS servers and including FPGAs, GPUs, and ASICs can support O-RAN Alliance RAN architecture.

4.7 Real-Time Intelligent Controller

The Real-Time Intelligent Controller is what truly makes the O-RAN architecture stand out compared to the traditional 3GPP architecture. The RIC is a cloud-native, software-defined component that enables the optimization and control of the RAN functionality improving network performance and allowing operators to adapt the RAN characteristics to suit each use case they offer to their customers. Leveraging artificial intelligence, machine learning, and open APIs, the RIC makes the RAN programmable. The RIC operators can onboard applications from multiple vendors while ensuring full interoperability with the other components of the RAN.

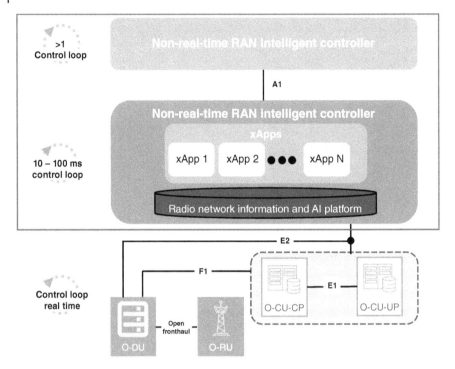

Figure 4.10 High-level architecture of the RIC. The Non-Real time RIC works on processes with long control loops (>1 s) while the near 0 Real Time RIC operates on processes with a 10–100 ms control loops.

Figure 4.10 shows a high-level diagram of the RIC. As it can be observed, the RIC consists of two components: the Non-Real-time RIC and the Near-Real-Time RIC.

4.7.1 Non-Real-Time RIC

4.7.1.1 Non-Real-Time RIC Architecture Principles

The Non-Real-Time RIC is part of the SMO and can interact with it as well as use the SMO interfaces such as O1 and O2, when required, to optimize the performance of the RAN. The Non-Real-Time RIC is dedicated to the control and management of the infrastructure. The attribute "Non-Real Time" reflects the fact that the control is applied to processes (Non-Real-Time functions) that that have cycles of over one second.

The functions managed by the Non-Real-Time RIC are called rApps and include analytics, configuration, and policy management which are generally delay-tolerant.

A further role of the Non-Real Time RIC is to manage the AI/ML models used in the Near Real Time RIC.

The Non-Real-Time RIC consists of two parts: the Non-RT RIC Framework, internal to the SMO that exposes the rApps via the R1 interface and the Non-RT RIC Applications (rApps) – modular applications that leverage the functionality exposed by the Non-RT RIC Framework to provide added value services relative to RAN operation.

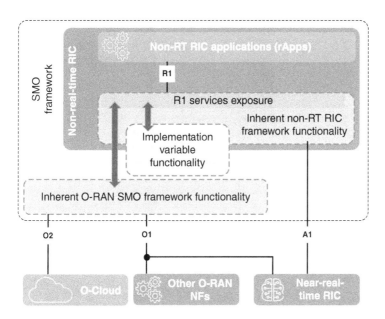

Figure 4.11 High-level architecture of the Non-Real-Time RIC. Source: Reproduced from O-RAN Alliance Adapted from [18].

4.7.1.2 R1 Services

The R1 services are provided over the R1 interface (see Figure 4.11) to offer the capabilities of the Non-RT RIC functional layers. Common services that can be provided include:

- Integration services, including service registration and discovery services, service authentication and authorization services and communication support services:
- Data management and exposure services.
- A1-related services.
- AI/ML workflow services.
- O1-related services.
- O2-related services.

4.7.2 Near-Real-Time RIC

The Near-Real-Time RIC [7] is a platform aiming to enhance the RRM functions of the RAN by means of multi-vendor programmable applications called xApps . The nomenclature "Near-Real-Time" derives from the fact that the control loop of the xApps is in the order of 10 ms (which is the duration of a radio frame in NR) or shorter. The Near-Real-Time RIC uses the E2 interface to connect to the RAN nodes it needs to interact with and is fully backward compatible with the 3GPP RRM meaning that it can control any RAN node that supports the E2 interface. A single Near-Real-Time RIC may be connected to several base stations.

Besides the E2, the Near-Real-Time RIC is also connected to the SMO via the O1 interface and can communicate with the Non-Real-Time RIC via the A1 interface and provides a layer for enabling APIs.

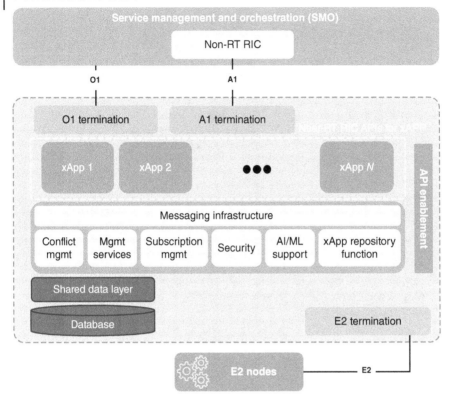

Figure 4.12 Architecture of Near-Real-Time RIC. Source: Reproduced from O-RAN Alliance Adapted from [7].

The architecture of the Near-Real-Time RIC is shown in Figure 4.12 and consists of the following components:

- *xAPPs*: The Near-Real-Time RIC can host several xAPPs, each designed to enhance a specific RRM function or a group of functions. The xApp has access to a wealth of RAN information coming from the RAN and generally uses a ML-trained model to calculate the most appropriate control command. As well as the database storing the RAN information, an xAPP can also leverage a common messaging infrastructure connecting the different parts of the platform. An xApp consists of a descriptor that includes the set of parameters needed to configure and manage the xApp and a software image which contains the actual xApp program.
- *Messaging infrastructure*: The messaging infrastructure provides the necessary functionality to allow different components of the Near-Real-Time RIC to communicate with each other. It supports registration, discovery, and subscription to events.
- *Conflict management*: The task of the conflict management component is to solve possible scenarios where two or more xApps attempt to issue control actions to the same target. For example, two xApps may attempt to set the same parameter to two different values or pre-allocate more resources than those available.
- *Subscription management*: xApps can use the messaging infrastructure to subscribe to events that are reported by the E2 Nodes connected to the Near-Real-Time RIC, and

the subscription management component optimizes the process by which xApps receive notifications and the data.

- *Security*: This component protects the RAN from potential malicious code in an xApp that may compromise the performance of the RAN.
- *AI/ML support*: Component that can be used by xApps to improve their efficiency.
- *xApp repository function*: This component ensures that the deployed xApps can be authenticated and authorized before they are enabled in the Near-Real-Time RIC and permitted to issue instructions to E2 Nodes. The xApp Repository Function can also be used for the discovery of xApps.
- *Shared data layer and database*: The database in the Near-Real-Time RIC consists of two parts: the RAN Network Information Database (R-NIB) that gathers RAN information coming from the connected E2 Nodes and a UE Network Information Base (UE-NIB) that contains information on the identity of the UE and enables xApps to track the UE across different E2 Nodes. To query the databases, a request is issued to the shared data layer.

Examples of functions that xApps can optimize include:

- load balancing based on UE control,
- radio bearer management,
- interference detection and mitigation.

The near-RT RIC enables both established telecom equipment manufacturers and new stakeholders to develop advanced algorithms that optimize the performance of the RAN taking into account the specific deployment scenario, the UE type and using machine learning to leverage the large amount of data available in the RAN.

4.8 Open Fronthaul

4.8.1 Addressing the Technical Challenges of the 5G Fronthaul

As seen in Chapter 3, in early distributed base station systems the RRH was mounted on a mast and connected via fiber or cable to the BBU which was normally located in a cabinet next to the same mast. It was therefore relatively straightforward to meet the latency and bandwidth requirements of the interface between these two elements. However, it was soon realized that for certain topologies (e.g. distributed antenna systems) and more importantly with the introduction of centralized RAN, it became necessary to create a more elaborate protocol for the interface between BBU and RRH which is referred to as fronthaul to distinguish it from the interface between the base station and the core network which is called backhaul. The requirement for the new protocol was to support the high data rate, delay-sensitive connection of the low physical layer with the antenna system.

The choice fell on the CPRI [20], but it was soon realized that CPRI suffered from two major problems:

- first, the interface bandwidth did not scale with the traffic but rather depended on the number of antennas. While early generations used relatively small bandwidth and not more than two antennas, this was not initially a big problem.

- second and probably more importantly, CPRI was not sufficiently tightly specified. This forced each vendor to apply proprietary extensions that resulted in the BBU to only work with selected RRH partners.

In 2017, the CPRI consortium released an evolution of the CPRI interface called eCPRI [21] that addressed the need for reducing the required bandwidth that 4G and especially 5G were going to need to connect a DU with an RU. While eCPRI introduced several enhancements including support for Ethernet as transport technology, cater for the higher bandwidth and low latency communications between RU and DU, many vendors were still compelled to apply proprietary extensions making it difficult for an operator to connect base station elements manufactured by different suppliers.

As the possibility of using different vendors has been the main tenet of the Open RAN philosophy, O-RAN Alliance, using eCPRI as a baseline, decided to define an open version of the fronthaul interface and to specify it to a level of detail sufficient to enable interoperability and avoid vendor lock-in. The result of this initiative was the release of the Open Fronthaul specifications [22]. The O-RAN Alliance Open Fronthaul is designed to cope with the typical 5G carrier bandwidth that could be in the order of 100 MHz, that is, five times the size of the typical 4G bandwidth, and include MIMO antenna systems that could contain tens of elements. While with CPRI the fronthaul interface bandwidth of such a 5G site could be in the order of several terabytes, with Open Fronthaul the bandwidth requirement drops to 10–50 GB (depending on the split applied) and the latency to around 1 ms: this level of performance is attainable with fiber links also over considerable distances.

The O-RAN Alliance Open Fronthaul consists of four distinct "planes":

- The C-Plane transfers real-time control between the O-DU and the O-RU.
- The U-Plane is used to transfer the actual user data, which when using the 7-2x split takes the form of IQ sample data, between the O-DU and O-RU.
- The S-Plane is used to ensure the tight latency requirements of the C-Plane and U-Plane are met in real time.
- The M-Plane takes care of non-delay-sensitive operations.

The eCPRI protocol was restricted to only define the U-Plane transport characteristics such as data frame, packet, and header formats, and it instead allowed for any protocol to be used to exchange control, synchronization, and management information. This was perceived as problematic in a multi-vendor environment, and as a consequence, O-RAN Alliance also strictly defines the additional traffic planes in their protocol architecture.

The protocol stack for the C-Plane and U-Plane is shown in Figure 4.13. As for eCPRI the Open Fronthaul relies on an Ethernet connection for the transport and on top of this, O-RAN Alliance allows for two variants of transport encapsulation: either encapsulation over Ethernet with VLAN or encapsulation over IP/UDP (User Datagram Protocol). In both cases, the packets containing C-Plane and U-Plane are marked in such a way as to receive the highest possible priority treatment.

Both eCPRI transport headers or Radio over Ethernet (RoE) or IEEE1914.3 [23], a protocol standardized by IEEE, can be used within the Ethernet payload according to O-RAN Alliance specifications. For each of the options, O-RAN Alliance defines a transport header providing information such as data flow type, sending and reception port identifiers,

Figure 4.13 Protocol stack of O-RAN Alliance fronthaul of U-Plane and C-Plane.

Section type: any								
0 (msb)	1	2	3	4	5	6	7 (lsb)	# of bytes
ecpriVersion				ecpriReserved			ecpriConcate nation	1	Octet 1
ecpriMessage								1	Octet 2
ecpriPayload								2	Octet 3
ecpriRtcid/ecpriPcid								2	Octet 5
ecpriSeqid								2	Octet 7

Figure 4.14 eCPRI transport header fields Adapted from [22].

whether concatenation of multiple application messages in a single Ethernet packet is supported, and so on. Both eCPRI and RoE can be carried over UDP (User Datagram Protocol).

For reference, the eCPRI and RoE transport headers are shown in the figures 4.14 and 4.15; it is however beyond the scope of this book to provide a detailed description of the fields therein for which the reader is referred to respective eCPRI definitions.

Figure 4.16 shows the S-Plane protocol stack defined by O-RAN Alliance [22]. As mentioned previously, this plane is used for frequency and time synchronization of O-DUs and O-RUs. The baseline architecture of this plane is the Synchronous Ethernet and the IEEE 1588 Precision Time Protocol (PTP) [24].

The Precision time Protocol can be transported directly when using L2 Ethernet (ITU-T G.8275.1 [25]) whilst when using UDP/IP (ITU-T G.8275.2 [26]) it is possible to only achieve partial timing support from the network and consequently a lower synchronization performance.

Section type: any								
0 (msb)	1	2	3	4	5	6	7 (lsb)	# of bytes
RoEsubType								1	Octet 1
RoEflowId								1	Octet 2
RoElength								2	Octet 3
RoEorderInfo								4	Octet 5

Figure 4.15 Radio over Ethernet transport header fields Adapted from [22].

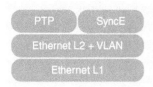

Figure 4.16 S-Plane protocol stack.

4.8.2 User Plane and Control Plane

As seen in the previous section, the U-Plane and the C-Plane share the same protocol stack, the same end points since and both connect the O-DU to the O-RU. They also share the characteristic of being very delay-sensitive, and for that reason there is no acknowledgment of the correct reception of a message, meaning that incorrect or lost messages cannot be recovered. It is common practice not to mix in the same message U-Plane and C-Plane messages.

The U-Plane transports the IQ samples containing the user's payload as well as come data that is used to improve the link performance.

In a similar manner as the M-Plane, the C-Plane messages transmit data-associated control information required for processing of user data, however. Unlike the M-Plane, the C-Plane operates in real time. The control of the downlink is performed independently from the control of the uplink using separate messages.

Examples of the procedures that require exchange of C-Plane messages include:

- Scheduling and beamforming commands
- Handling of physical random access channel and mixed numerology
- Download precoding configuration parameters and indications
- Symbol numbering and duration
- Dynamic spectrum sharing (DSS)
- Channel information

4.8.3 Synchronization Plane

Synchronization across the different elements of the O-DU and O-RU is required, especially in those situations where a UE may be receiving downlink data from multiple RUs simultaneously, since this way the throughput can be maximized. Synchronization is also needed during handover from one RU to another following which happens as the UE moves around the network. The accuracy of the synchronization is heavily influenced by the network implementation and network topology; for this reason, O-RAN Alliance considers the following synchronization options that can be used over an Ethernet network:

- Frequency synchronization where clocks are aligned in frequency.
- Phase synchronization where clocks are aligned in phase.
- Time synchronization where clocks are aligned to a common base time.

Figure 4.17 M-Plane protocol stack.

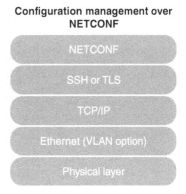

Configuration management over NETCONF

- NETCONF
- SSH or TLS
- TCP/IP
- Ethernet (VLAN option)
- Physical layer

4.8.4 Management Plane

The M-Plane is defined by O-RAN Alliance in a dedicated technical specification [27]. The role of the M-Plane is to carry non-delay-sensitive messages between that enables the O-DU to prepare and manage the O-RUs connected to it. For example, the O-DU can send to the O-RU radio parameters such as frequency band, MIMO settings as well as information to configure the fronthaul.

The operations carried out over the M-Plane include the following:

- "startup" installation,
- software management,
- configuration management,
- performance management
- fault management and
- File management.

The protocol stack of the management plane is shown in Figure 4.17.

The chosen network management protocol is NETCONF as defined in RFC 6241 [28] and the data modeling language YANG as defined in RFC 7950 [29]. The rationale for this choice is that, besides NETCONF/YANG being a common combination widely adopted in the industry, it allows easy interworking between products with different capabilities as well as an architecture whereby multiple nodes can be configured simultaneously. The network manager protocol is carried using a secure layer (SSH or TLS) on top of TCP/IP.

References

1 ITU-R, "Recommendation ITU-R M.2012-5, Detailed Specifications of the Terrestrial Radio Interfaces of the International Mobile Telecommunications – Advanced (IMT Advanced)," 2012.

2 GSMA, "5G Implementation Guidelines," 2019.

3 3GPP, "TS 37.470 W1 Interface General Aspects and Principles," 2014.

4 3GPP, "TS 36.423 X2 Application Protocol (X2AP)," 2010.

5 O-RAN Alliance, "O-RAN Architecture Description v06.00," 2022.

6 O-RAN Alliance, "Cloud Platform Reference Designs," 2021.

7 O-RAN Alliance, "Near-Real-time RAN Intelligent Controller, Near-Real-time Architecture," 2021.

8 O-RAN Alliance, "A1 interface: General Aspects and Principles v2.03," 2021.

9 O-RAN Alliance, "E2 interface: General Aspects and Principles v2.01," 2021.

10 O-RAN Alliance, "O1 Interface Specification for O-CU-UP and O-CU-CP," 2021.

11 O-RAN Alliance, "O1 Interface Specification for O-DU," 2021.

12 O-RAN Alliance, "O2 interface: General Aspects and Principles v1.02," 2022.

13 O-RAN Alliance, "Control, User and Synchronization Plane Specification v8.00," 2022.

14 3GPP, "TR 38.801 "Study on New Radio Access Technology: Radio Access Architecture and Interfaces," 2017.

15 "Wikipedia," [Online]. Available: using the common Open System Interconnection (OSI) model. [Accessed 2023].

16 3GPP, "TS 38.300 "NR; NR and NG-RAN Overall Description, Stage 2," 2023.

17 C. Aleksandra, H. Lehrmann, and L. M. P. Larsen, "A survey of the functional splits proposed for 5G mobile crosshaul networks," *IEEE Communications Surveys & Tutorials*, vol. 21, no. 1, pp. 146-172, 2019.

18 O-RAN Alliance, "O-RAN Non-RT RIC Architecture," 2021.

19 O-RAN Alliance, "O-RAN Cloud Architecture and Deployment Scenarios for O-RAN Virtualized RAN," 2022.

20 CPRI, "Common Public Radio Interface (CPRI); Interface Specification," 2013.

21 CPRI, "Common Public Radio Interface: Requirements for the eCPRI Transport Network," 2018.

22 O-RAN Alliance, "Cooperative Transport Interface Transport Control Plane Specification v2.00," 2021.

23 IEEE, "IEEE Draft Standard for Radio over Ethernet Encapsulations and Mappings," 2019.

24 IEEE, "IEEE Standard for Local and Metropolitan Area Networks – Time-Sensitive Networking for Fronthaul," 2018.

25 ITU-T, "G8275.1/Y.1369.1 (2020) Amendment 2 (06/2021) Precision Time Protocol Telecom Profile for Phase/Time Synchronization with Full Timing Support from the Network," 2021.

26 ITU-T, "G8275.2/Y.1369.2 (2020) Amendment 2 (06/2021) Precision Time Protocol Telecom Profile for Time/Phase Synchronization with Partial Timing Support from the Network," 2021.

27 O-RAN Alliance, "Management Plane Specification v8.00," 2022.

28 IETF, "RFC 6241, "Network Configuration Protocol (NETCONF)"," 2011.

29 IETF, "RFC 7950, "The YANG 1.1 Data Modeling Language"," 2016.

5

TIP – Commercialization of Open RAN

5.1 Overview

As described in Chapter 2, TIP's objective is focused on the commercialization of Open RAN, and therefore TIP's activities in this area merit attention. TIP approaches the commercialization of Open RAN in the concept of releases. A release consists of three major stages (note that there can be six steps as shown in Figure 5.1, but they have been grouped into three stages for easier comprehension).

First, the use cases and technical requirements of the product are defined based on the 3GPP standards and O-RAN profiles. Different features of standards are prioritized and selected with productization in mind. Then, the list of features and operator requirements is circulated to the vendors of Open RAN, where they are asked to respond which features will be supported in different timeframes (e.g. 6, 12, and 18 months). These responses then form the industry roadmap that the release of the product will be based on.

Second, the product built by the group is tested and validated such that the product is proven to be ready for field deployment. To achieve this, TIP defines the test and validation requirements that are independent of the different deployment scenarios that operators desire. This is to ensure that the difference in deployment scenarios will not impact the performance of the product, and the product delivers consistent quality regardless of the region being tested (and hopefully deployed). In the end, a badge is awarded, depending on the level of testing environment, to the tested product.

Finally, the product is released in the TIP Exchange, where the product becomes available for any interested player to deploy and the product blueprint becomes available. The product blueprint is important because it highlights sets of products that have been integrated and tested together [1]. In this final stage, the product can be individual network nodes (RU/DU/CU) and the product blueprint can be a set of RU, DU, and CU tested together as an integrated network.

In a more detailed view, a release will consist of the following steps, where the steps will be repeated as more releases are introduced.

- Gather inputs from MNOs.
- Assess technology supplier compliance.
- Initiate TIP roadmapping process.
- Publish TIP OpenRAN Release definition.

Open RAN Explained: The New Era of Radio Networks, First Edition.
Jyrki T. J. Penttinen, Michele Zarri, and Dongwook Kim.

Figure 5.1 Three stages of TIP Release. Source: Adapted from [1].

- Publish TIP OpenRAN product requirements.
- Issue requirements-compliant ribbons for technology suppliers.
- Publish product test plans and issue supplier-validated product badges.
- Publish blueprint definitions and blueprint test plans.
- Initiate test and validation in TIP community labs and third-party authorized labs.
- Issue TIP-validated product badges and publish validated blueprints on TIP Exchange.

5.2 Fundamental: Requirements and Test Plans

5.2.1 OpenRAN Overview

In terms of setting the requirements and test plans, the OpenRAN is the PG (Project Group) in charge of TIP. It is aimed at building disaggregated RAN solutions, where it focuses on the development of technical reference implementations under industry collaboration and on the commercialization of the implementations [3].

As of January 2023, TIP's OpenRAN PG has touched upon and addressed issues in the following areas [2]:

- Disaggregation of RAN HW and SW on vendor-neutral, General-Purpose Processor (GPP)-based platforms.
- *Open interfaces*: implementations using open interface specifications between components (e.g. RU/CU/DU/RIC) with vendor-neutral hardware and software.
- Multiple architecture options, including:
  - o An all-integrated RAN with disaggregation at SW and HW level
  - o A split RAN with RU and BBU (DU/CU)
  - o A split RAN with RU, DU, and CU.
  - o A split RAN with integrated RU/DU and CU.
- *Flexibility*: multi-vendor solutions enabling a diverse ecosystem for the operators to choose for their 2G/3G/4G and 5G deployments.
- Solutions implemented on bare metal, virtualized platform, or containerized platform.
- Adoption of novel technologies (e.g. AI/ML).

5.2.2 OpenRAN Releases

5.2.2.1 Overview of OpenRAN Releases

As of August 2022, there are three OpenRAN releases: 0, 1, and 2. Release 0 was focused on the macrocell in the outdoor settings for 2G, 3G, and 4G. Starting with a technical requirements document, commercial products built by Parallel Wireless, a participating vendor, were tested in Vodafone's network in Turkey. The products were based on integrated RU-BBU architecture with disaggregation of hardware and software. After the trial and subsequent testing, products were awarded with badges/ribbons and were listed on TIP Exchange [2, 4].

Release 1 expands the scope in many ways: the products defined support 5G, the components developed are more fine-grained, and indoor is supported in addition to outdoor settings. This means that Release 1 is expected to provide offerings that will comprise full (or level equivalent to full) RAN. Component-wise, Release 1 defines the RU, the DU and CU, the Radio Intelligence and Automation, and RAN orchestration and management. The components are intended to be integrated under O-RAN 7.2 specifications-based architecture. The testing and validation stage is still ongoing with planned live trials around the world [2, 4].

Release 2 is the first release based on the updated OpenRAN MoU Group (OMG) technical priorities document, where the OMG is led by Deutsche Telekom, Orange, Telefonica, Telecom Italia Mobile (TIM), and Vodafone. While the initial version of the document was published in June 2021, the operators updated the requirement in March 2022, with a focus on the intelligence, orchestration, transport, and cloud infrastructure applied to the RAN. Note that not all the features will be included from day 1 but will be divided into a series of minor releases 2.x as shown in Table 5.1. Therefore, some features will be implemented in H1 2022 as part of Release 2.0, some in H2 2022 as Release 2.1, and the rest in H1 2023 as Release 2.2. Once the latest Release 2.x is stable, the Release 2 products will be scaled for commercial operation [2, 4].

These minor releases of Release 2 differ in terms of the deployment scenarios and features supported (see Table 5.1). Namely, Release 2.0 consists of the "basic" deployment scenarios and features for the typical RAN. For example, it supports the macro and indoor cells along with both standalone and non-standalone deployments. It also supports the basic features of SMO, RIC, and Self-Organizing Network (SON). Furthermore, positioning and mobility features are supported in Release 2.0. Release 2.1 is characterized by enhanced features and network sharing. Whereas Release 2.0 is focused on single-operator network, Release 2.1 introduces sharing scenarios. It also supports mmWave and advanced features such as network slicing and VoNR (NOTE: most appropriate term for this is Voice over 5G System, Vo5GS). Finally, Release 2.2 brings in additional deployment scenarios such as the IoT scenarios and also enhancements of the features introduced in Release 2.0.

Release 3 builds on the previous releases of OMG RAN technical priorities document. The OMG published its updated requirements in March 2023, where it focuses on the further requirements on SMO and RIC along with enhanced nodes such as the cloud infrastructure, O-CU/O-DU, and O-RU. It is also worth noting that there have been improvements in the topics related to security and energy efficiency [5].

Table 5.1 Release 2 Roadmap: deployment scenarios and features.

Release 2.0 (1H 2022)	Release 2.1 (2H 2022)	Release 2.2 (1H 2023)
• Macro-distributed • Macro-centralized • Legacy RAT support • Cloud infra • Indoor mono operator • Indoor small cell • SON basic • SA • NSA • Positioning • Mobility • Capacity and dimensioning – basic • SMO and automation basic • Carrier aggregation intra-site • Dual connectivity • SON ANR • Dynamic spectrum sharing • QoS • RIC basic	• RAN sharing/MORAN distributed • Single-vendor MORAN with open fronthaul • Multi-vendor RAN sharing/RAN sharing MORAN centralized • RAN sharing/MOCN • RAN sharing management • Carrier aggregation inter-site • mmWave support • SON advanced • RIC basic • Slicing • VoNR	• NB-IoT • LTE-IoT • Indoor multi-operator • RIC advanced • SMO advanced • Energy efficiency

Source: TIP OpenRAN Release 2 [4] Telecom Infra Project (TIP).

As can be seen above, publicly available information mostly centers around Release 2 and only OMG RAN technical priorities documents are public for Release 3. Therefore, this section will describe Release 2 roadmap and progress along with the highest priority requirements (with unanimous P0 priority) of OMG as of Release 3 in order to provide a complete picture of the latest OpenRAN work.

5.2.2.2 RU

Different features of RU are supported in different minor releases (see Table 5.2) of Release 2. In Release 2.0, basic features for operation are included such as the massive MIMO and 7-2x interface. In Release 2.1, mmWave is supported to align with deployment scenarios and features. Finally, IoT scenarios are supported in Release 2.2.

Table 5.2 Release 2 Roadmap: RU.

Release 2.0 (1H 2022)	Release 2.1 (2H 2022)	Release 2.2 (1H 2023)
• Single-, dual-, and tri-band RUs • mMIMO with 8T8R 32T32R • Low, mid, and high band support • 2T2R, 4T4R, and 2T4R macro RUs • Support of O-RAN-compliant 7-2x interface with O-DU	• Carrier aggregation • mm Wave support	• Dual connectivity • Carrier aggregation • NB-IoT • LTE – IoT

Source: TIP OpenRAN Release 2 [4] Telecom Infra Project (TIP).

In Release 3, the OMG operators consider the support of an embedded physical sensor inside of the O-RU for measuring and reporting energy efficiency metrics a "must have." In addition, the O-RU shall support 8 : 2 Time Division Duplex (TDD) slot pattern configuration and adaptation of the uplink waveform depending on the circumstances. The O-RU must be able to handle a certain number of MIMO layers (4 for RRHs and 16/8 for massive MIMO antennas).

5.2.2.3 DU/CU

DU/CU follow similar patterns (see Table 5.3). Release 2.0 consists of the basic features for a single provider of the RAN. The DU/CU supports key interfaces and protocol stacks for basic operation, while the underlying cloud platform supports implementation on commodity hardware and basic management functions. In Release 2.1, the DU/CU supports safety/reliability features for commercial operation and the cloud platform is enhanced to support the full functionalities of the basic functions. Finally, in Release 2.2, the DU/CU supports multi-provider scenarios.

Table 5.3 Release 2 Roadmap: DU and CU.

Release 2.0 (1H 2022)	Release 2.1 (2H 2022)	Release 2.2 (1H 2023)
DU/CU 2.0 • DU/CU Hardware Platforms Aligned with HW Requirements – Cell Site DU/CU, Edge Cloud DU/CU, Data Center CU • Synchronization with IEEE1588v2 and SynE • IPv4, IPv6, and Dual Stack IPv6/IPv6 • IPSec for Midhaul and Backhaul • Support of DU/CU Containerization • Single Provider F1, X2/Xn, and E1 interface • CU Connected to Multiple DUs **O-Cloud Infra** • Bare Metal Implementation • Board Management • Acceleration, Crypto, and Network Drivers • Support of RTOS • Synchronization of GPS, PTP, and SyncE • Basic Support of AAL, O-Cloud Management, and Security • Kubernetes Plugin	**DU/CU 2.0** • Full Support of External Alarms of DU/CU • Cellsite Outdoor Temp −40 to 55 °C • Scalabilities of Containerized vDU, vCU • Multi-Provider F1 interface (indoor) • Inter gNB-DU Mobility for EN-DC • O-Cloud Implementation **O-Cloud Infra** • Full Support AAL, Configuration Management, Host Management, Security • Full Support of O-Cloud Management, e.g. BM Provisioning, Underlay Network Provisioning • Multi-Cluster Workload Lifecycle, Configuration and Policy Management	**DU/CU 2.0** • Multi-Provider F1 interface (Macro) • Open W1 interface • Multi-Provider E1 interface • DU Connectivity to Multiple CU-CPs • Intra gNB-DU CA • Centralized retransmission in Intra gNB-CU scenarios

Source: TIP OpenRAN Release 2 [4] Telecom Infra Project (TIP).

In Release 3, DU and CU will support the mobility of cells within one O-DU. That is, the continuity of service is ensured for UE moving among the cells of the O-DU. The most prominent requirement for DU and CU is the introduction of Advance Configuration and Power Interface (ACPI) and mechanisms to manage energy efficiency. For example, CPUs will be able to enjoy the flexibility of different power state configurations based on the load/circumstances. Artificial traffic can be generated to measure power consumption for different traffic levels as defined in ETSI ES 202 706-1.

5.2.2.4 RIA

In TIP, RIA stands for RAN Intelligence Automation and is usually concerned with the RIC. The roadmap for RIA is shown in Table 5.4. In Release 2.0, basic features of the non-RT RIC and near-RT RIC are supported. In Release 2.1, more complicated use cases such as quality optimization and machine learning application are supported. Finally, sharing and slicing scenarios are supported in Release 2.2.

In Release 3, there are major updates for both non-RT RIC and near-RT RIC. In addition to alignment with O-RAN alliance interfaces such as A1 and R1, the non-RT RIC now needs to support interaction with external platforms/services (e.g. oversight platforms and AI/ML platforms). The non-RT RIC also needs to allow consumers of data to register data types they consume/produce and to subscribe/request registered data types, allowing exposure and management of data in non-RT RIC.

Near-RT RIC is similar to non-RT RIC in that there is an alignment with O-RAN alliance specifications on near-RT RIC. The most notable requirement within the alignment is

Table 5.4 Release 2 Roadmap: RIA (RIC).

Release 2.0 (1H 2022)	Release 2.1 (2H 2022)	Release 2.2 (1H 2023)
• Traffic Steering detailed requirements from RIA R2 aligned with OMG • Energy efficiency detailed requirements from RIA R4 aligned with OMG • Basic non-RT RIC platform requirements to use cases defined in Rel 2.0 • Basic RI platform requirements for rAPPs support • Test plans for use cases • Test results from vendor testing	• mMIMO detailed requirements from RIA R1 and X1 aligned with OMG • QoE optimization detailed requirements created by RIA «RX» aligned with OMG • QoS-based resource optimization detailed requirements created by RIA «Rx» aligned with OMG • RIC platform requirements for A1 policy, A1 enrichment, and A1 ML deployment • RIC platform requirements for rAPPs support • Conflict management • RIC platform requirements for xAPPs support	• RAN sharing detailed requirements created by RIA «Rx» aligned with OMG • RAN slice SLA assurance detailed requirements created by RIA «Rx» aligned with OMG • Enhanced RIC platform requirements for previously supported functionality

Source: TIP OpenRAN Release 2 [4] Telecom Infra Project (TIP).

that the near-RT RIC needs to support hosting third-party xApps and not limited to the ones developed by the platform vendor. This indicates disaggregation within the near-RT RIC, where it allows the RIC platform vendor and xApp provider to be different player.

5.2.2.5 SMO

Similar to the case of RIA, SMO in TIP is usually covered in ROMA (RAN Orchestration and Management), whose roadmap is shown in Table 5.5. Release 2.0 of ROMA is based on the ETSI MANO architecture and supports basic features such as the functionalities of O1 interface. Release 2.1 brings service-based architecture into play and adds support for O2 interface along with other features. Finally, ROMA is enhanced to support additional features of different interfaces and network slicing in Release 2.2.

SMO is also the most impacted and upgraded entity in Release 3. SMO shall support the management of license, where the utilization of different software licenses can be monitored and potentially generate an alarm if there is an excess usage beyond a certain threshold. SMO also needs to support enhanced "plug and play" of Open RAN objects (e.g. infrastructure, network functions, slices, xApps/rApps, etc.), meaning that registration, auto-discovery, and auto-update of these objects need to be supported.

Table 5.5 Release 2 Roadmap: SMO (RAN Orchestration and Management).

Release 2.0 (1H 2022)	Release 2.1 (2H 2022)	Release 2.2 (1H 2023)
• ETSI MANO-based reference architecture • Support OMG deployment scenario 2, 3, 14, 15, 18, and 19 • RAN NF can be virtualized or centralized • Basic support of O1 i/f for FCP management • Provider-specific i/f between ROMA components • Provider-specific northbound APIs for ROMA service exposure • LCM of network services and NFs • Alarm handling and correlation of alarms from cloud infra and NFs • Basic performance monitoring and reporting • Configuration management and zero-touch provisioning, backup, and restore • Automated CD/CT workflow engine	• Service-based reference architecture • Enhanced support of O1 i/f for FCP management • Basic support of 3GPP/ORAN i/f between ROMA components • Basic support of standards-based northbound APIs for ROMA service exposure • Auto-healing and fault root cause analysis • Enhanced performance monitoring and reporting • Basic support of O2	• Support OMG deployment scenario 4&5 • Support RAN network slicing • Advanced support of O1 i/f for FCP management • Enhanced support of O2, and 3GPP/ORAN i/f between ROMA components • Enhanced support of standard-based Northbound APIs for ROMA service exposure • Enhanced auto-healing, fault root cause analysis, and fault mitigation • Support RIC deployment • Support non-RT RIC and ROMA interaction

Source: TIP OpenRAN Release 2 [4] Telecom Infra Project (TIP).

In addition, the principles regarding AI and ML modeling have been adopted in Release 3. The principles include fair, reproducible, explainable, auditable, monitorable, and support of continuous training for AI/ML models.

5.3 Testing and Validation, Marketplace

5.3.1 Rationale

The rationale of testing and validation and marketplace is to adapt the procurement model of operators in the era of open paradigm [6]. Traditionally, operators had to approach vendors bilaterally to work out the features and profile for the type of deployment they have. This can be feasible for a few incumbent vendors, but the complexity increases when the number of vendors becomes large. Not only will the operators require more resources for communication, but also the vendors are faced with significant burden in communication.

This is where TIP comes into play with the OpenRAN PG. Each release centralizes and combines as much as possible of the baseline product requirements that a vendor needs to support. It also provides a baseline level of testing based on those requirements. This helps operators save time, resources, and costs as some of the integration and system-level testing are already done within TIP. Vendors benefit by getting an idea of what features need to be prioritized on the roadmap and also by eliminating the need to produce different variants for different operators.

While the requirements and test plans form the basis of this centralization, testing and validation process, along with marketplace, is the ultimate stage where such value is ensured for operators and vendors.

5.3.2 The Process

Traditionally, each vendor was tasked to perform a full RAN system validation for the operator [6]. This has an advantage in that the operator can be concerned less with the integration of the system and can manage the network system by communicating with a few (if not one) selected vendors. Indeed, this paradigm works for monolithic deployments even in the 5G era.

In the Open RAN paradigm where there will be many vendors, this is no longer feasible. It is very resource-consuming for an operator to integrate many vendors and also challenging to communicate with these vendors. This means that testing and integration of the large number of vendors can happen only a few times a year, whereas having one vendor as the integrator and having limited number of components allow easier testing to ensure performance. Further complicating the issue is that the Open RAN technology testing of the community is based on plugfests, which are usually hosted only once or twice a year.

This calls for a process relying on continuous testing of components and releases. For example, automating the testing process enables delivering the same level of testing as was the case for the traditional approach. This is where TIP's Innovation Hub comes in. It is a federated fulfillment model with "Test once; Deploy many times" approach. This model focuses on a number of different blueprints and configurations, where the tests are done in a centralized mechanism. Note that it is at this stage where best practices are shared and field trials are encouraged among the TIP community members.

Now, based on the defined blueprints, configurations, and tests, the badges are awarded to participating suppliers depending on the level of testing. There are three types of badges: bronze, silver, and gold. Bronze badges demonstrate a vendor's self-assessed compliance with TIP requirements, while silver and gold badges involve increasing levels of testing. The vendors are then listed on the TIP Exchange for anyone to see the type of badge the vendors have won.

Bronze badge is awarded when a vendor responds to the list of product requirements. This implies that the vendor is request for comments (RFI) ready and is on the TIP Exchange, enabling operators/customers to access this information. Having a bronze badge indicates that the vendor satisfies a common industry RFI created by TIP, and the operator can benefit in two ways. First, in the case of the operator having similar RFI, the process can be skipped as TIP has already done it. Second, even in the case when the operator wants to issue a different RFI, the information can still help in the process.

Silver and gold badges are awarded when a vendor satisfies test requirements and/or test plans defined by TIP's PGs. Silver is awarded when a product is validated in the TIP community lab or an approved third-party lab, while gold is awarded when a solution is validated in an end-to-end environment.

Note that a bronze or silver badge does not mean that a particular product, combination, or solution is not suitable for deployment. Rather, it merely means that the operators will typically require more in-house testing before deploying a bronze or silver product, combination, or solution. For example, Tier 1 operators with enough resources may continue relying on their own testing facilities and use bronze- or silver-badged products. The smaller operators, on the other hand, may prefer silver or gold-badged solutions to avoid in-house testing.

5.3.3 Achievements

TIP has made a few achievements in this area of testing and validation and marketplace. First, TIP has successfully set up a testing environment for key Open RAN differentiator technologies. For example, the orchestration and management automation product/solution [7]. It is established in the TIP community lab located in Meta's Menlo Park Campus and consists of Foxconn's RU and Capgemini's CU/DU and 5GC. Another example is the RAN Intelligence and Automation (RIA), which involves the RIC. TIP established a testing lab for the RIA group in Meta's Menlo Park and BT's Adastral Park, where O-RAN Software Community RIC, VMWare RIC, Accelleran RIC, and VIAVI RIC Tests are supported [8].

Second, it has promoted trials and deployments in relation to its testing and validation process. As shown in Figure 5.2, there are many markets that have trialed and

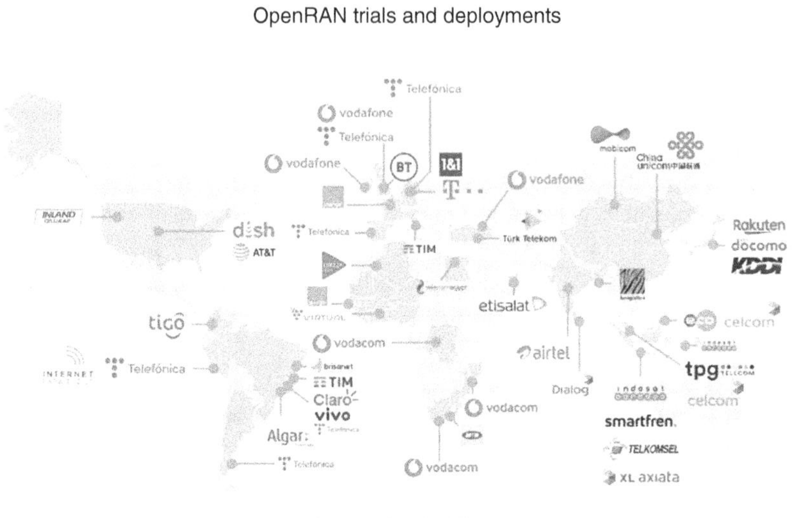

OpenRAN trials and deployments

Figure 5.2 OpenRAN Trials and Deployments (as of Oct 2022). Source: Data taken from TIP OpenRAN PG accelerates Open RAN Commercialization.

deployed the Open RAN technology. Some of the trials have resulted in a playbook (e.g. Vodafone Turkey's Open RAN playbook [9]) to share the best practices and promote deployments.

5.4 Experience of OpenCellular

To understand the final objectives of TIP and to get an idea of what the future of Open RAN under TIP can look like, it is beneficial to note the experience of OpenCellular, a graduated[1] PG at TIP. This is because OpenCellular is a similar, although much smaller in scale, initiative as OpenRAN, and was geared toward opening up the details of OpenCellular network equipment (GSM and LTE) such that anyone can produce and operate them.

The OpenCellular PG's mission was to "connect the unconnected by empowering communities with tools to build and operate sustainable community cellular infrastructure, using open-source technologies and an open ecosystem." Therefore, it focused on developing the following with mostly rural deployments in mind:

1. *Hardware*: to develop open-source cellular infrastructure hardware that can be utilized by various types of "customers" (e.g. MNO, community operators, academic, and researchers).
2. *Software*: to enable carrier-grade and open-source end-to-end solution tailored for the needs of the communities without royalty.
3. *Ecosystem and platform*: to foster service ecosystems to support development, deployments, operations and maintenance, and to build local human capacity.

---

1 At TIP, a concluded PG is denoted as graduated.

4. *Fostering deployment*: to help the deployment of community cellular networks using grant programs and partnering with other global/regional initiatives with similar objectives of promoting connectivity.

With OpenCellular, the customers were able to expand into areas with low population density, where high total cost of ownership (TCO) had previously deterred them from expanding cellular coverage. Especially taking into account that OpenCellular provided open-source, publicly available files that are optimized for rural deployments (e.g. reduction of power consumption), there was a potential to reduce TCO significantly [10].

On the other hand, the vendors were able to benefit from OpenCellular as the development of solutions was shared and even the smaller vendors can reduce the resources required to acquire a certain level of performance. In addition, publicly available hardware designs and features allowed the vendors to develop new versions, allowing flexibility to adapt to different markets [10].

The OpenCellular also partnered with other entities and partners to promote the deployment and operation. As local entities were involved to build and provide support in deployment, the customers were able to reduce equipment import and labor costs. Involvement of more entities also meant that iteration of applying the design and making improvements that are contributed to the community was more frequent, improving the design [10].

It is no surprise that there were various community networks and service providers in Africa and Latin America using OpenCellular. In 2020, Africa Mobile Networks installed sites based on OpenCellular covering over 300,000 people in the African region, providing coverage for MTN and Orange.

References

1 Telecom Infra Project, "TIP: Facilitating, Accelerating and Enabling Open RAN," 16 01 2023. [Online]. Available: https://telecominfraproject.com/tip-facilitating-accelerating-and-enabling-open-ran/. [Accessed 10 05 2023].

2 ABI Research, "The Road to Open RAN Productization," 2021.

3 Heavy Reading, "TIP OpenRAN: Toward Disaggregated Mobile Networking," 2020.

4 Telecom Infra Project, "TIP OpenRAN Release 2 Roadmap," 2022.

5 Telecom Infra Project, "OpenRAN MOU Group," 2023. [Online]. Available: https://telecominfraproject.com/openran-mou-group/#release3. [Accessed 29 August 2023].

6 Telecom Infra Project, "Testing & Validation Process for Open RAN Success," 16 03 2023. [Online]. Available: https://telecominfraproject.com/testing-validation-process-for-open-ran-success/. [Accessed 10 05 2023].

7 Telecom Infra Project, "OpenRAN Orchestration Lab Testing Successful Completion," 08 02 2023. [Online]. Available: https://telecominfraproject.com/openran-orchestration-lab-testing-successful-completion/. [Accessed 10 05 2023].

8 Telecom Infra Project, "TIP RIC-Rolls OpenRAN innovation with RAN Intelligence and Automation (RIA) Group," 08 02 2023. [Online]. Available: https://telecominfraproject

.com/tip-ric-rolls-openran-innovation-with-ran-intelligence-and-automation-ria-group/. [Accessed 10 05 2023].

9 Telecom Infra Project, "Playbook – OpenRAN Trials with Vodafone Turkey," 2020.

10 Telecom Infra Project, "Open Cellular - Wireless Access Platform Design," [Online]. Available: https://telecominfraproject.com/opencellular/. [Accessed 10 05 2023].

6

Open RAN Use Cases

6.1   Introduction

To be successful commercially, a technology per se does not suffice on its own, but it needs to be complemented by use cases and their continued evolution. Open RAN does not only enable invigoration of the RAN ecosystem, but its new features bring into reality new use cases and better support for traditional use cases. Especially, the ability to leverage intelligence in the control of the RAN and the ability to program the RAN is the key differentiator of Open RAN.

In this context, this section will describe the use cases of Open RAN, which can be grouped into the ones that serve as enabling foundations and the others that serve the use cases of specific domains. The former set includes those that enhance customer experience, optimize network deployment, and enable advanced RAN capabilities such as slicing and sharing. The latter is where the Open RAN is applied to specific domains such as the connected mobility and private network deployments of vertical businesses.

6.2   Open RAN as Enabling Foundation

6.2.1   Overview

This section describes the use cases that makes O-RAN serve as the enabling foundation to be applied to business context [1]. These can be grouped into four categories. First, there are use cases that enhance and maintain customer experience. Second, there are use cases that facilitate the component technologies of 5G networks. The third group of use cases enable network slicing in the RAN in multi-vendor configuration. Finally, there are use cases that enable more advanced network sharing capabilities and lead to the concept of network-as-a-service (NaaS). It should be noted that the third and the last group of use cases have some overlap, but are not identical.

6.2.2   Customer Experience

6.2.2.1   QoE Optimization

The O-RAN architecture, with its intelligence capabilities, supports real-time quality of experience (QoE) optimization. While today's quality of service (QoS) framework is

Open RAN Explained: The New Era of Radio Networks, First Edition.
Jyrki T. J. Penttinen, Michele Zarri, and Dongwook Kim.
© 2024 John Wiley & Sons Ltd. Published 2024 by John Wiley & Sons Ltd.

sufficient to satisfy the requirements of less demanding 5G use cases such as mobile broadband and sensor networks, providing demanding 5G use cases such as cloud virtual reality (VR) and cloud gaming requires predictive capability to ensure end-user satisfaction even amid uncertain radio conditions.

This use case is realized by new interfaces and nodes defined by the O-RAN architecture. First, the non-RT RIC constructs and trains the artificial intelligence (AI)/machine learning (ML) models based on the QoE-related measurement metrics from network-level measurement report and service management and orchestration (SMO). The models are then deployed in near-RT RIC, where the non-RT RIC continues to send policies/intents. Second, the near-RT RIC takes the model constructed, trained, and updated by the near-RT RIC and executes the model to optimize QoE. The near-RT RIC also sends QoE performance report to non-RT RIC for evaluation and optimization. Third, the RAN nodes communicate via the O1 and E2 interfaces. The O1 interface is used between SMO and open distributed unit (O-DU) to report network state and UE performance, while the E2 interface is used between near-RT RIC and RAN nodes (O-CU, O-DU, and O-eNB) to enforce QoS (O-CU refers to open centralized unit, and O-eNB refers to open eNB).

At high level, the QoE optimization use case flows as follows. For the AI/ML model training, the non-RT RIC collects measurements and data from the SMO (and possibly application server) to develop new ML model or modify existing ML model. The model is then trained and deployed in the near-RT RIC via O1 interface. The models are assigned for different xApps in the near-RT RIC via A1 interface. For the case of policy generation, the non-RT RIC collects the relevant data, evaluates them, and generates the appropriate QoE optimization policy. This is then sent to near-RT RIC via A1 interface, where the near-RT RIC makes inference on the RAN conditions based on the policy and the data collected from other RAN nodes. The inference leads to corresponding policy or control message that is sent to the RAN nodes via E2 interface. Note that for model update and policy generation, the performance of existing model/policy can take place by non-RT RIC (model and A1 policy) and near-RT RIC (E2 policy), where they collect relevant data and update accordingly.

6.2.2.2 Signaling Storm Protection

In addition to QoE optimization, signaling storm protection is another use case that is related to maintaining customer experience. This use case attempts to protect both intentional and accidental signaling attacks to the mobile network by identifying and mitigating the signaling storms as quickly and locally as possible (see Figure 6.1). This will ensure reliable operation of the mobile network.

Similar to QoE optimization, the near-RT RIC, non-RT RIC, and new interfaces defined by O-RAN enable this use case. The non-RT RIC maintains a global view and therefore detects the signaling storms that can be taking place over wide range of operations and not confined to a specific location. Data from other sources (e.g. 5G core) can be used to assist with the global view and more accurate detection/classification of the attacks. Of course, the non-RT RIC also pushes out relevant AI/ML models and policies to the near-RT RICs to detect and respond to signaling anomalies. The near-RT RIC contains various xApps

1. Malicious device sends multiple signals
2. Many devices **coincidentally** send signals concurrently

Figure 6.1   Signaling storm illustrated.

(e.g. signaling storm detection and signaling storm mitigation) to detect abnormal levels of signaling on the E2 interface and enforce policies. The near-RT RIC also takes the AI/ML model and the policy from the non-RT RIC to monitor and detect signaling anomalies. The RAN nodes communicate with the near-RT RIC via E2 interface such that the detection and the policy enforcement can take place.

The high-level flows for the signaling storm protection proceed as follows. The first high-level flow corresponds to the case where the near-RT RIC detects the local signaling storm and applies corresponding policies. Here, the signaling storm is detected by the signaling storm detection xApp in near-RT RIC from the signaling messages report it receives from the RAN nodes over E2 interface. When a UE is detected to be aggressive, the signaling storm detection xApp updates the signaling storm mitigation xApp, which creates filter to block/throttle signaling messages from the malicious UE. This is then applied to the RAN nodes via E2 interface, where the RAN nodes evaluate the policy and rejects/throttles signaling from the malicious UE. Depending on the share of UEs that behave aggressively, the policy filter can be updated accordingly. Optionally, the decision on restricting the malicious UE can take place in the near-RT RIC instead of the RAN nodes, and this is called the local signaling storm protection insert control.

Another high-level flow corresponds to the case where the non-RT RIC detects the network-level signaling storm attacks and applies stricter configuration to near-RT RIC to mitigate the attack. This flow relies on the OAM functions to collect, process, and send related metrics/information to the non-RT RIC. The non-RT RIC uses AI/ML model to analyze the received information, which is then applied to in the near-RT RICs. When the signaling storm activity originating from various locations or with uncorrelated identifiers is detected, the non-RT RIC updates the configuration in the near-RT RICs to be stricter, where the near-RT RIC performs detection and mitigation as they would for the local signaling storm protection. When the management information indicates that the signaling storm is no longer present, the non-RT RIC restores the configuration of near-RT RIC to the initial setting.

6.2.3 Facilitating 5G Component Technologies

6.2.3.1 Overview
The O-RAN architecture also comes with enhancements that can facilitate key 5G component technologies. For instance, 5G comes with massive MIMO (mMIMO) configuration that enables beamforming, and the O-RAN architecture can optimize beamforming in 5G especially in multinode/multi-vendor scenarios. In addition, the potential of steering of traffic in the NG-RAN, dynamic spectrum sharing (DSS), local indoor positioning, and energy saving can be maximized with Open RAN.

6.2.3.2 Massive MIMO Beamforming Optimization
5G comes with mMIMO technology, and this enables beamforming where multiple antenna can be configured to form a beam toward the user. However, the treatment and optimization of mMIMO require careful consideration as it can be deployed in different scenarios (e.g. macro cell only vs. macro and 3D small cells), and it may also come with vendor specific fixed system parameters. This means that the mMIMO system is either too complicated to configure flexibly (i.e. too many parameters and scenarios to consider) or the system is reliable but is not flexible for the operator to configure.

To overcome these, the O-RAN architecture utilizes the SMO, non-RT RIC, and near-RT RIC to bring more intelligent and flexible approach to optimize mMIMO. Here, the SMO plays a central role, as it collects the necessary information (configurations, measurements, performance indicators, etc.) and configures the optimized beam parameters for RAN nodes and/or near-RT RIC. The non-RT RIC is involved in the mMIMO grid of beams (GoB) forming optimization, where it collects information (including the ones from the network and the applications) from SMO and the RAN to construct/train the relevant AI/ML models, where the models are used to provide the optimized beam pattern configuration for SMO to configure the RAN nodes (see left side of Figure 6.2 for concept of GoB).

The near-RT RIC is involved in the beam-based mobility robustness optimization (MRO), and it collects data from the RAN nodes to construct/train the relevant AI/ML models. The inference is used to create optimal configuration for parameters related to beam mobility that is consistent with the global optimization configuration from the SMO. For example, as indicated in Figure 6.2, handover should not be triggered even if the right cell has stronger signal because the device is better served under the coverage of the left cell. In other words,

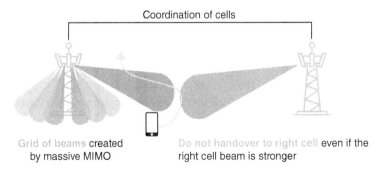

Grid of beams **created**
by massive MIMO

Do not handover to right cell **even if the**
right cell beam is stronger

Figure 6.2 Concept of GoB and MRO illustrated. Source: [2, 3].

the optimization ensures that the service quality due to mobility is disrupted minimally and manages the beams accordingly.

In both cases of non-RT RIC and near-RT RIC, the performance of the model is monitored and retrained if necessary. Finally, the RAN nodes collect and send KPIs of near-RT RIC related to beamforming and beam mobility to SMO or near-RT RIC. Note that they also apply the configuration set by the SMO and near-RT RIC.

Within beamforming optimization, there are two high-level flows. The first involves non-RT RIC and is concerned with mMIMO GoB beamforming optimization. Here, the objective is to optimize multicell mMIMO beamforming flexibly and to allow auto-mated/intelligent solution. To do so, the non-RT RIC collects data from the SMO, the RAN, and relevant applications to train the AI/ML model. Then, the model inference is used to derive optimal mMIMO configuration parameters, which is then forwarded to SMO such that SMO can configure the RAN accordingly. The second is where the near-RT RIC is involved and focuses on the MRO aspect of mMIMO beamforming. The near-RT RIC collects data from the RAN nodes and trains the relevant AI/ML model. The inference then is used to configure the RAN nodes, which will be optimized to ensure beam mobility and to comply with the operator's target beamforming optimization.

Note that the performance of the optimization will be evaluated for both cases, where necessary modifications to the AI/ML model will be made.

6.2.3.3 MU-MIMO Optimization

Further building on the mMIMO technology, the presence of multiple antennas allows the cell to form multiple beams, each targeted at a specific user (see Figure 6.3). Since the beams are formed toward respective users, there is less interference and potentially provides higher cell capacity along with better performance for users. This is the multiuser MIMO (MU-MIMO) use case, but there is yet another challenge of having to deal with different mobility patterns of the users. For example, some may be stationary, some may be moving at a low speed, and others moving at a high speed. The intelligence of the Open RAN can be utilized to resolve this challenge.

In the MU-MIMO use case, the near-RT RIC is the key actor. It communicates with the RAN nodes (specifically, E2 nodes such as the O-CU and O-DU) to first collect the rele-vant information such as the UE state, cell channel quality, and cell configuration. The near-RT RIC, based on this information, provides recommendations to the RAN nodes on

Figure 6.3 MU-MIMO illustrated. Source: Adapted from [4].

the optimal configuration for accommodating the UEs in the cell. The RAN nodes, on the other hand, provide relevant information and reports to the near-RT RIC and apply the configurations provided by the near-RT RIC.

The high-level flow of the MU-MIMO use case starts with the near-RT RIC triggering the collection of relevant information (cell configuration, UE states, traffic, and channel information) from the RAN nodes. The RAN nodes respond and provide the near-RT RIC with the requested information. This is not one-time event but is periodically done for the near-RT RIC to accumulate data. Then, the near-RT RIC uses the information collected to calculate the optimal configuration (e.g. in which frequency and time resources should a particular UE be allocated to) and sends the optimized parameters to the RAN nodes. The RAN nodes then implement the recommendations provided by the near-RT RIC.

6.2.3.4 Traffic Steering

Another key 5G component technology facilitated by O-RAN architecture is the multi-RAT dual connectivity (MR-DC). In MR-DC, the network needs to identify where a given traffic needs to be steered toward and to ensure the QoS is satisfied across different types of traffic/multiple RAT (see Figure 6.4). For example, in NR E-UTRA dual connectivity (NE-DC) setting, should a given traffic be diverted to the NR master node through master cell group (MCG) bearer or secondary cell group (SCG) bearer? Or should it utilize the LTE secondary

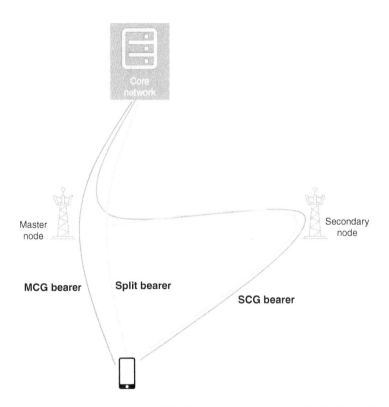

Figure 6.4 Traffic steering in MR-DC (only including case with master node).

node? How would this impact the performance of the RAN overall? In this context, more network wide view of traffic steering and more proactive (rather than responsive) management of traffic will be beneficial for mobile network optimization.

In traffic steering use case, the non-RT RIC and the near-RT RIC are the key actors. The non-RT RIC collects the necessary performance, configuration, and measurement data so that policies and measurement configurations can be defined and updated for traffic steering. These are then communicated to the near-RT RIC and the RAN nodes. The near-RT RIC takes the policy from non-RT RIC and enforces them. The RAN nodes are responsible for collecting and sending the necessary data to the SMO.

The high-level flow of traffic steering starts with the non-RT RIC's action to define and communicate the traffic steering policies to near-RT RIC. Examples include preferences on which cells to allocate control/user plane and handling of bearer. Then, near-RT RIC interprets and enforces the policy given by the non-RT RIC. The monitoring of the data continues throughout the process, and when the traffic steering policy is no longer deemed necessary, the policy is deleted.

6.2.3.5 Dynamic Spectrum Sharing

The third key 5G component technology of this subsection is DSS. DSS is where the mobile operator can dynamically allocate radio resources across its 4G and 5G access networks, enhancing the spectrum utilization and efficiency overall. As shown in Figure 6.5, the resource can be dynamically allocated and can prioritize particular radio access if necessary.

The support for DSS in the context of Open RAN is to accommodate multi-vendor scenarios and further network optimization (e.g. traffic prediction and resource management). This is enabled by the RICs and can be abstracted as shown in Figure 6.6, where both non-RT RIC and near-RT RIC control and coordinate the 4G and 5G RAN nodes to dynamically allocate spectrum resources.

In this use case, the non-RT RIC translates SMO's service requirements specific to DSS into resource sharing policies which can be applied to the RAN. Consequently, it provides long-term policies for 4G and 5G as well as policies on what actions should be taken if performance deviates from expected KPIs. The non-RT RIC, as usual, is responsible for developing and training AI/ML models for the near-RT RIC and retrain these models based on the policy feedback from near-RT RIC if necessary.

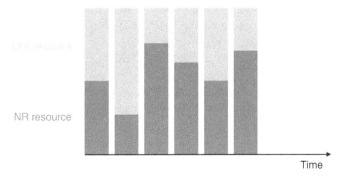

Figure 6.5 Concept of dynamic spectrum sharing. Source: Adapted from [5].

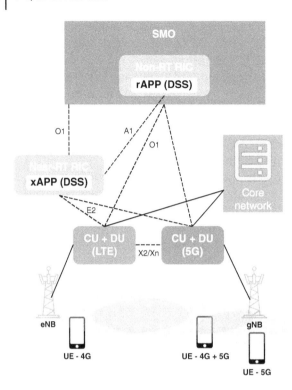

Figure 6.6 RIC-based DSS architecture. Source: Adapted from [1]. (Not all use cases and related depicted functions and terms are implemented in the current O-RAN specifications).

The near-RT RIC takes the model and policy from non-RT RIC in the form of ML model and xApps and interprets the policies related to resource allocation per different radio access technologies (RATs). Furthermore, it translates the RAT-specific policies to be applied to other RAN nodes (especially the ones connected via E2 interface: O-CU and O-DU) and provides feedback to non-RT RIC on the model and the policy.

The relevant RAN nodes take part by supporting the following functionalities over E2 interface. It receives the DSS-related configuration, supports resource-management-related metric collection, and accepts control and policy enforcement from near-RT RIC. In addition, they share collected data with non-RT RIC over O1 interface.

High-level flow of DSS using O-RAN proceeds as follows. First, the SMO sets DSS policies collects and deploys it in the non-RT RIC. SMO also collects DSS-related data from near-RT RIC and O-RAN nodes of 4G and 5G. The data is then forwarded to non-RT RIC, where non-RT RIC develops and trains the model. The model is published by the SMO which is then deployed in near-RT RIC. The non-RT RIC deploys relevant DSS policies in near-RT RIC. The near-RT RIC then collects RAN configuration and operation parameters from the E2 nodes, and based on these, the near-RT RIC now provides configuration parameters to the E2 nodes for DSS. During the operation, the E2 nodes provide report on their performance/scheduling and status, where the near-RT RIC executes the ML model and modifies configuration if required. Finally, near-RT RIC provides policy feedback to non-RT RIC, and the SMO provides DSS-related data to non-RT RIC that it collected. This prompts updates in policies and the model by non-RT RIC if necessary.

6.2.3.6 Local Indoor Positioning

The ability to derive indoor position of the UE in real time based on the cellular network measurements enable applications by not only the mobile operators but also other industries. 3GPP Release 16 already supports positioning capability of 5G NR by introducing the location management function (LMF) within the NG-RAN, but this feature involves the AMF of 5GC and sometimes may require long routing of messages between NG-RAN and LMF. Open RAN introduces the possibility to calculate and extract the position information locally within RAN, which reduces the delay and makes the information timely.

In this local indoor positioning use case, the key entities are non-RT RIC and near-RT RIC. Near-RT RIC is the crucial component, as it selects relevant positioning algorithms and carries out the calculation of the actual indoor position. Non-RT RIC can enhance the process by collecting relevant network-level measurement report, building and training the AI/ML model to assist in position calculation of near-RT RIC, and sending relevant policies to near-RT RIC to optimize the positioning information further. Other RAN nodes, such as the O-CU and O-DU, provide the necessary information to the RICs.

The high-level flow of the local indoor positioning use case can take place in two scenarios. The first scenario only involves the near-RT RIC and the RAN nodes. In this scenario, the external application server requests the RAN for positioning information of a particular UE. The near-RT RIC receives the request and requests the relevant information from the RAN nodes. Based on the responses of the RAN nodes, the near-RT RIC calculates the positioning of the UE, which is then notified to the external application server.

The second scenario involves non-RT RIC and SMO in addition to the entities in the first scenario. Here, the start is the same with external application server making a request to near-RT RIC, but involves the non-RT RIC and SMO in leveraging accumulated information and positioning calculation AL/ML model. Provided that the non-RT RIC and SMO have already been collecting information and building and training the AI/ML model, the non-RT RIC provides the AI/ML model to near-RT RIC upon request of the external application server. The near-RT RIC again asks for information from the Open RAN nodes, but this time uses the AI/ML model given by non-RT RIC in addition to own algorithm in calculating the position. The result is again notified to the external application server, and it is also possible for near-RT RIC to provide feedback on the positioning result derived by given AI/ML model.

6.2.3.7 Energy Saving

RAN consumes majority of the energy that the mobile network consumes, and in the era of climate change and sustainability, it is imperative for the networks to enhance its energy efficiency to achieve more with less energy and consequently carbon dioxide emissions. While there are many solutions being developed to save energy of the mobile network, a solution that is worth covering is the dynamic switching of carriers and cells depending on the circumstances at the time of writing. The Open RAN can optimize this solution by applying its intelligence capabilities.

In the energy savings use case, the entities involved depend on the two types of deployment. The first type of deployment is where the AI/ML inference host resides in the non-RT RIC. Here, non-RT RIC, E2 interface nodes (e.g. O-CU and O-DU), and the O-RU are involved. Non-RT RIC is in charge of collecting relevant information and performance reports from the SMO, E2 nodes, and O-RU for training and inference of the AI/ML model to be used. It also deploys and trains/retrains the AI/ML model. The E2 nodes report the relevant information and performance reports to the non-RT RIC via SMO with O1 interface and implement actions required for energy savings. O-RU is similar to E2 nodes in that it implements required actions and reports the relevant information, but the reporting can be done via O-DU or via SMO.

In this first type of deployment, the high-level flow proceeds as follows. First, the SMO collects the measurements data from E2 nodes and O-RU. This data is then provided to the non-RT RIC for AI/ML model training and inference. Then, the non-RT RIC uses its AI/ML model and forwards to deduced actions (cell/carrier switch off/on) to SMO, which instructs the E2 nodes and the O-RU (for this, E2 nodes can intermediate or SMO can instruct directly) accordingly. The non-RT RIC continues to monitor and analyze the performance of the AI/ML model, where it can decide to update/retrain the model.

The second type of deployment for energy savings use case is where the AI/ML inference host is in the near-RT RIC instead of non-RT RIC. In this case, we have an addition of near-RT RIC to the first type of deployment. In this case, non-RT RIC and near-RT RIC divides the labor, where non-RT RIC trains, deploys, and configures the AI/ML model in near-RT RIC. The near-RT RIC analyzes the received data from E2 nodes and provides deduced policies/optimization toward the E2 nodes. Furthermore, the E2 nodes can communicate directly with near-RT RIC instead of going through SMO in reporting relevant information.

In this deployment, the high-level flow proceeds as follows. The start is the same where the SMO collects relevant data from the E2 nodes and the O-RU. The data is again provided to the non-RT RIC for AI/ML model training and inference. However, the deployment of the model is done in the near-RT RIC. Furthermore, the SMO triggers and provides target for the energy optimization, and the non-RT RIC may provide policies for near-RT RIC. The near-RT RIC then collects data from the E2 nodes and the O-RU, where the deployment model is used to deduce actions, and the actions are then implemented in E2 nodes and the O-RU.

6.2.4 Network Slicing

6.2.4.1 Overview
The O-RAN architecture also focuses on supporting novel features in 5G, where network slicing is one of the key 5G features supported by O-RAN. O-RAN Alliance suggests that with O-RAN, four use cases related to network slicing can be achieved: QoS-based resource optimization, RAN slice SLA assurance, multi-vendor slice, and NSSI resource allocation optimization.

6.2.4.2 QoS-Based Resource Optimization
Guaranteeing the QoS for a slice with preferential treatment end-to-end is an important use case to fulfill for 5G networks where network slices are used to differentiate different types of customers and use cases. In this case, isolating resource between slices and monitoring the slice service quality is essential to achieve the preferential treatment. In the context of RAN, the physical resource blocks (PRBs) are isolated between slices

such that QoS requirements can be met in the RAN while optimizing the performance of the slices overall. While these are done based on rigorous planning and field testing, there are always exceptions and circumstances that make the configured resource or PRB isolation inadequate for fulfilling the QoS. This is where the intelligence and management functionality of the Open RAN, via SMO and the RICs, can improve the situation. For example, it can prioritize or deprioritize particular users of the slice such that QoS of the slice can be met. Or it can choose to allocate more PRBs and resources to the preferential slice to meet the desired QoS level.

In this QoS-based resource optimization use case, the SMO, non-RT RIC, near-RT RIC, and the E2 nodes are involved. The SMO collects relevant data, the non-RT RIC monitors necessary metric related to QoS of different network functions and provides policies to near-RT RIC, the near-RT RIC interprets and enforces the received policy from non-RT RIC, and the E2 nodes provide relevant data to the SMO and enforces QoS based on the messages from the near-RT RIC.

The high-level flow of the QoS-based resource optimization use case proceeds as follows. First, the SMO collects relevant data from the E2 nodes, and the non-RT RIC monitors and evaluates the measurements. When congestion is detected, the non-RT RIC requests additional measurements to SMO for more detailed evaluation and requests additional information on priorities of the users of the slice to the customer (external server) via SMO. Based on this information, the non-RT RIC creates policy and forwards the policy to the near-RT RIC. The near-RT RIC enforces the policy in the E2 nodes. The measurement of the E2 nodes and the evaluation of the situation by non-RT RIC continue. When the congestion is resolved, the policy can be deleted.

6.2.4.3 RAN Slice Service Level Assurance

The network slices are not just defined in terms of QoS, but there are many aspects of the service they support that need to be fulfilled. The provision of these aspects is based on the service level assurance (SLA) between the network operator and the customer. Ensuring the SLA of a slice end-to-end and especially in the context of RAN is critical to providing network slicing in a manner that satisfies the customer. The Open RAN can enable the network to ensure SLA for each slice and to prevent possible violations by utilizing its open interfaces and AI/ML-based intelligence capabilities.

This RAN slice SLA assurance use case involves various entities of the Open RAN. The SMO collects relevant data from the E2 nodes, the near-RT RIC, and business customers (i.e. external application/server), as well as providing the RAN slice SLA. The non-RT RIC monitors RAN slice performance, trains the AI/ML models or creates the control apps for optimization, and creates relevant policies and reconfiguration requests to be provided to different entities depending on the optimization scenario. The near-RT RIC monitors in near-real-time RAN slice performance, deploys and executes AI/ML models or control apps depending on the optimization scenario, and performs actions to assure RAN slice SLA via necessary means. Finally, the E2 nodes will provide slice-specific performance measurements/reports and will implement slice-specific actions that will assure the RAN slice SLA.

The high-level flow of RAN slice SLA assurance use case can be decomposed into three components, where the first is creation and deployment of the SLA assurance models and apps, and the other two are slow/fast loop optimization of the RAN Slice SLA. The creation and deployment of the SLA assurance models and apps proceeds as follows. First, the non-RT RIC retrieves RAN slice SLA from the SMO, and the SMO provides the data it collected from the RAN and (if required) the external application to the non-RT RIC. This information is then used to train AL/ML models. Depending on the use of the AI/ML models and the control apps, the AI/ML models or the control apps are deployed in non-RT RIC or near-RT RIC depending on the optimization scenario. The non-RT RIC continues to monitor the measurements and reports and can choose to retrain/update the AI/ML model or to update the control app.

The flow for the optimization to assure RAN slice SLA depends on whether the optimization is slow loop or fast loop. In slow loop, the AI/ML model or the control app resides in the non-RT RIC, and the SMO updates slice configuration of near-RT RIC and the E2 nodes. Therefore, it starts when the non-RT RIC decides that either the RAN should be reconfigured or slice-specific A1 policies should be created. In any case, the collected information via O1 interface will be used, accompanied by slice SLA requirements, operator-defined intents, and enrichment information if necessary. If it is reconfiguration, the non-RT RIC will ask the SMO to update the slice configuration of near-RT RIC and the E2 nodes, which the SMO will then deliver the updates. If it is policy update, the non-RT RIC will send the updated policies to near-RT RIC for its enforcement on E2 nodes. Finally, the near-RT RIC and the E2 nodes, regardless of being reconfiguration or policy update, will execute and enforce the received updates and assure SLA.

In fast loop, the AI/ML model or the control app is deployed in the near-RT RIC. In this case, non-RT RIC generates policy based on collected information and various criteria such as the SLA requirements and the enrichment information from the external server. The near-RT RIC then receives the policy from non-RT RIC and slice-specific configuration information from SMO. It also starts collecting data from the E2 nodes. Based on the measurements and evaluation, the near-RT RIC starts SLA assurance for the RAN slice and enforces the policy/configuration on E2 nodes.

6.2.4.4 Multi-vendor Slice

The next network slicing use case is multi-vendor slice, which refers to the case where the network slice can consist of vO-DU and vO-CU functions from different vendors (see Figure 6.7). Since network slicing is a key 5G feature, being able to provide this with

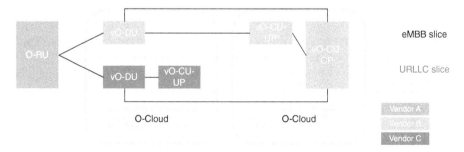

Figure 6.7 Multi-vendor slice example (Core network entities not shown). Source: Adapted from [1]. (Not all use cases and related depicted functions and terms are implemented in the current O-RAN specifications).

different vendors in an open manner is important in developing Open RAN. The O-RAN Alliance also highlights three benefits associated with multi-vendor slice.

- More flexibility and faster time to market: the operator can enjoy a pool of multiple vendors along with less time to deploy a particular feature/function in the network.
- RAN sharing cases facilitated with greater flexibility: the operators do not have to have own physical equipment in sharing scenarios, but simply share one physical equipment.
- Reduced supply chain risk: reduce dependance on individual vendors.

In this use case, the O-RAN alliance proposes a solution with the following entities involved: the multi-vendor slice app in SMO, near-RT RIC, and E2 nodes. The multi-vendor slice app configures the virtual O-DU and O-CU functions, as well as configuring O-RU to connect to virtual O-DU functions. The near-RT RIC shares UE-unique data (e.g. MAC) among virtual O-DU functions and communicates configuration parameters to RAN. The E2 nodes (physical O-RU and virtual functions of O-DU and O-CU) deal with traditional RAN functions, but with division of labor. The virtual function of O-DU that acts as primary vO-DU processes signaling radio bearer (SRB), data radio bearer (DRB), and other vO-DU-related functions, while the secondary vO-DU function processes only DRB-related functions.

O-RAN Alliance provides two scenarios in where the solution can be useful. One is having multi-vendor slice over one operator, and another is accommodating two operators in RAN sharing. In the first scenario, slice #1 is mapped to the primary vO-DU and vO-CU functions and slice #2 is mapped to the secondary vO-DU and vO-CU functions. O-RU is shared between the primary and the secondary vO-DUs, and the control plane of CU (CU-CP) is shared between the user planes of the primary and the secondary vCUs (CU-UP). In high level, the UE registers with the primary virtual functions, where the near-RT RIC function relays relevant information (e.g. UE ID, cell ID, and share information) to the secondary vO-DU. Then, the UE starts PDU session establishment procedure for slice #1 with the primary virtual functions, where session establishment procedure for slice #2 is initiated and the secondary vO-DU is requested to setup and modify UE context. Then, the allocation of resources is determined by the RIC. Allocation to primary vO-DU means data transmission over slice #1 and to secondary vO-DU means data transmission over slice #2.

In the second scenario, PLMN #1 and #2 replace slice #1 and #2 of the first scenario, respectively. Therefore, PLMN #1 is mapped to the primary vO-DU and vO-CU functions, and PLMN #2 is mapped to the secondary vO-DU and vO-CU functions. In addition, O-RU is shared between the primary and the secondary vO-DUs, and the control plane of CU (CU-CP) is shared between the user planes of the primary and the secondary vCUs (CU-UP). The case where the UE wants to access PLMN #1 is trivial, as it would resemble the normal registration procedure in traditional RAN. Therefore, only the case where the UE wants to access PLMN #2 is described. In high level, the UE attempts to register to PLMN #2, where the request is first received by primary virtual network functions, and then relayed by the near-RT RIC function to the secondary vO-DU. Then, the UE establishes PDU session with the secondary virtual network functions. Finally, the RIC allocates resources to secondary vO-DU and enables the UE to make data transmission to PLMN #2 via the secondary virtual network functions.

6.2.4.5 NSSI Resource Allocation Optimization

The final network slicing use case is network slice subnet instance (NSSI) resource allocation optimization. As 5G comes with diverse use cases ranging from traditional broadband services to sensor networks, it is important for the RAN to be able to accommodate the diverse use cases in a flexible manner. In the management plane, the network slices for these diverse use cases are supported by the concept of network slice instance (NSI) and the NSSI, where NSSI for the O-RAN nodes can be optimized to address the different service requirements (see Figure 6.8). The resource allocation to different NSSIs in Open RAN can be optimized by AI/ML models, trained based on large dataset of the past usage patterns. Furthermore, policy for the NSSIs can prioritize resources of more premium slices in the times of congestion.

In this use case, SMO, non-RT RIC, and RAN nodes are involved. The SMO provisions in advance the default NSSI resource quota policy, which may be used by non-RT RIC when the relevant information is required (e.g. was not specified when the slice was created). The non-RT RIC collects the measurements and information from the RAN nodes, trains AI/ML models based on the historical measurements to predict traffic demand patterns of different NSSIs at different times and locations, determines resource allocation based on the AI/ML model, and updates configurations to optimize the NSSI resource allocation. The RAN nodes provide the required measurement and implement the updates in configuration.

The high-level flow proceeds as follows. First, the non-RT RIC builds the AI/ML model required to predict traffic demand patterns of different NSSIs. Then, it collects measurements and information from the RAN nodes, where this information are used to make inference of the AI/ML model. The inference is then translated into actions by non-RT RIC, which will be either reconfiguration of NSSI attributes or request to the O-Cloud management and orchestration function in order to update the O-Cloud resources used by

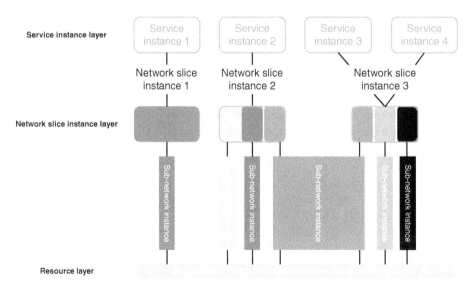

Figure 6.8 NSI and NSSI in realizing network slicing. Source: Fabrizio Granelli [6].

the network slice. In the case of the former, the RAN nodes will implement the updated configuration to optimize resource allocation.

6.2.5 Network as a Service and RAN Sharing

6.2.5.1 NaaS: Concept and Deployment

The support for network slicing naturally leads to the possibility of providing network as a service (NaaS). NaaS is a business model and ultimately an operating entity that build, own, and operate dedicated networks, which can be utilized by service operators on subscription or on-demand basis.[1] In other words, the end users of service operators subscribing to the NaaS infrastructure would connect to the infrastructure of the NaaS operator and then be routed to the service operators' infrastructure of billing/operations support system depending on the subscription (i.e. some service operators may choose to subscribe to only part of the network infrastructure and use their own infrastructure in other parts). The end users would pay the service operators, but a part of it will be shared with NaaS operator as subscription fee. These considerations are summarized in Figure 6.9.

As can be seen from the concept, Open RAN is not a prerequisite to NaaS, and it can be deployed based on the traditional networks. However, the Open RAN can facilitate NaaS by opening interfaces and disaggregating layers. For example, by opening interfaces, it becomes easier to consolidate existing network/infrastructure into one network. In addition, disaggregation of layers motivate the involved stakeholders to share hardware resources to achieve economies of scale in the hardware infrastructure.

TIP has been progressing works on NaaS in the context of open networking. It ranges from the playbook on NaaS to detailed guidelines for testing and site economics [7]. Since

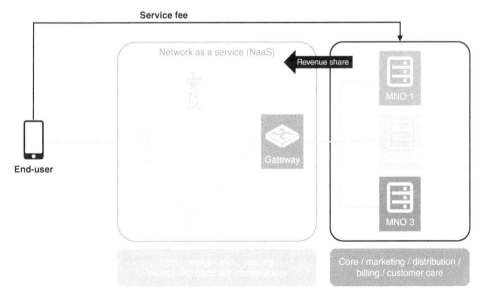

Figure 6.9 The concept of NaaS. Source: Adapted from [7].

---

1 Modified from the definition of TIP.

these are mostly focused on guiding deployment and implementations, the contents of TIP's work will be covered in Chapter 8.

6.2.5.2 Application to RAN Sharing

NaaS is in fact not a new concept, but it has already been a feasible solution in the industry for a long time. Sharing of network and infrastructure is a form of NaaS, and sharing of RAN should be discussed. Consequently, O-RAN Alliance has identified key use cases of Open RAN that can facilitate the sharing of RAN among different entities.

The first use case is to support multioperator RAN (MORAN), which is a configuration where the RAN is shared among the operators, but dedicated spectrum is used by each sharing operator [8]. In MORAN, O-RAN Alliance is interested in providing Open RAN solution that can improve the flexibility in configuration of each sharing operator, whereas in the past, the configuration was constrained by strategy and goals of other sharing operators.

This use case focuses on the case of geographically split RAN, where each operator deploys network and infrastructure in the regions it is responsible for. The use case proposes to have the responsible operator's infrastructure and computing resources available for hosting other operators' virtual network functions. Note that RIC nodes are not shared, but each operator's RIC node controls its virtual units in remote sites via remote E2 interface and communicates the desired configurations to counterpart RICs via O1 and O2 interfaces. The two-operator scenario is depicted in Figure 6.10.

Since this use case considers geographical split, it would be symmetric to consider the network functions of one operator hosted in the RAN site of another operator. Therefore, we focus on the case of having network functions of operator B in the RAN site of operator A for convenience here.

In this case, we have the SMOs of both operators A and B, the infrastructure management framework (IMF) of operator A, the RAN of the site of operator A, and the non-RT and

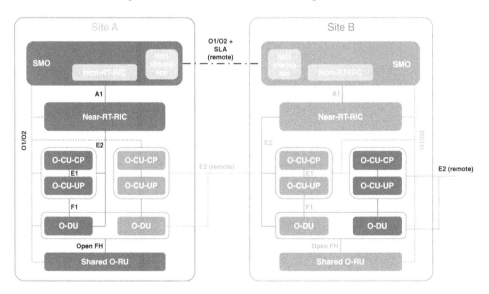

Figure 6.10 Open RAN MORAN use case architecture. Source: Adapted from [1]. (Not all use cases and related depicted functions and terms are implemented in the current O-RAN specifications).

near-RT RICs of operator B involved. From the operator B, we have the SMO that sends the commands (remotely since they are physically separate) via O1/O2 interfaces remotely[2] to deploy and configure operator B's VNFs in operator A site. These are sent to the sharing APP in the SMO of operator A. The SMO of operator B also forward RAN-related data (e.g. UE KPIs) collected by operator B's VNFs at the site of operator A to the non-RT RIC of operator B. The non-RT RIC of operator B configures the initial network policy template of operator B's VNFs at the site of operator A. It also sends policy/intention to these VNFs via near-RT RIC. The near-RT RIC of operator B is responsible for monitoring and collecting radio state report from the operator B's VNFs at the site of operator A via E2 interface remotely.

At operator A, the SMO checks the orchestration and management requests sent by the SMO of operator B is within the agreed SLA. Then, it asks the IMF to deploy operator B's VNFs, which are then configured further based on the requests of operator B sent via O1/O2 interfaces. It also forward the RAN-related data collected by the operator B's VNFs in the site of operator A to the SMO of operator A. The IMF consequently creates VNFs of operator B based on the request of SMO of operator A, and the RAN of the site of operator A collects relevant data and conveys them to operator A via E2 or O1/O2 interface remotely.

The high-level flow of RAN sharing proceeds as follows. First, the sharing App of operator B's SMO requests the sharing App of operator A's SMO to provision necessary VNFs at the site of operator A. The sharing App of operator A's SMO verifies that the request is in line with the SLA and then requests the IMF of cloud platform infrastructure of the site of operator A to instantiate the corresponding VNFs of operator B via orchestrator within SMO. The IMF then creates these VNFs and notifies the sharing App of operator A's SMO, which is then conveyed to the sharing App of operator B's SMO and to its orchestrator. The SMO of operator B will then proceed with configuration, where the orchestrator and the sharing App of operator B will send configuration request to the sharing App of operator A's SMO. The sharing App will again verify that the configuration is within the SLA and configures the VNFs of operator B at the site of operator A via orchestrator. During operation of the VNFs, the near-RT RIC of operator B is subscribed to the reports of VNFs at the site of operator A via E2 interface remotely. The VNFs will send relevant reports (e.g. radio state report and UE KPIs) to the near-RT RIC of operator B via E2 interface remotely and to the non-RT RIC of operator B via O1/O2 interfaces remotely. At this point, the non-RT RIC may decide to update network policy/intent based on the collected data. If this is the case, it will forward the updates to near-RT RIC of operator B, which will then update accordingly the configuration of VNFs of operator B at the site of operator A.

6.2.5.3 Sharing of O-RUs

The RAN sharing use case does not only include the scenario where O-DU and O-CU are shared but also includes sharing of O-RU. However, the O-RAN Alliance has provided the sharing of O-RU as an independent use case because the O-RU is shared among multiple O-DUs. This means that the use case can also be used for single operators for purposes such as resiliency and load balancing. Nevertheless, it can be generalized to the case involving multiple operators and therefore may include configuration systems of more than one operator. The solution presented by O-RAN alliance requires the use of inventory management

---

2 The term "remote" is used to emphasize that it does not refer to a direct connection between the two nodes, but the connection takes place through operator A's SMOs and/or other intermediate nodes.

systems that identify available RAN nodes (O-RU and O-DUs) for sharing and partitions O-RU resources statically between O-DUs.

The use case can be decomposed into sub use cases, where each of them serves function(s) to enable sharing of O-RU. The first group of sub use cases focuses on getting shared O-RU operational and includes resource partitioning, start-up, configuration, and supervision sub use cases. The second group of sub use cases focuses on maintaining the service of the shared O-RU and includes supervision, performance, resiliency, and antenna line device (ALD) control.

The first step for enabling shared O-RU is about resource partitioning. Here, the goal is for the rApp of the SMO of the shared O-RU operator to acquire details on how shared O-RU's resource is to be partitioned. The sharing coordinator decides to share an O-RU between multiple O-DUs and recovers the inventory of O-RU and O-DUs, then the O-RU resource is partitioned and shared with the rAPP in the SMO. The rAPP then communicates the involved O-DU identities to configuration management system, which will configure the transport systems accordingly.

The second step takes care of the start-up of the O-RU. After resource partitioning, the O-RU synchronizes with the O-DUs that are to share the O-RU. Then, depending on the hybrid management mode or hierarchical management mode, the shared O-RU calls home to SMO (hybrid) or to the O-DUs (hierarchical) in order to establish network management session. Then, each entity (SMO or O-DUs) will be contacted to check for software updates and install and activate the software in O-RU if necessary.

The third step configures the O-RU, which follows after the start-up and software update of the O-RU. The configuration also takes the management mode that was taken during the start-up. The shared O-RU orchestration rAPP at the SMO will trigger the configuration of the common aspects of the shared O-RU at non-RT RIC, which will also trigger the configuration. For hierarchical management (call home to O-DUs), the rApp will determine which O-DU is responsible for the common configuration before the SMO starts triggering. Then, the SMO will use OpenFronthaul interface to configure common aspects of shared O-RU in hybrid (or neutral host) management mode and use O1 interface via O-DU in the hierarchical mode [9]. In the hierarchical mode, O-DU will use OpenFronthaul interface to make the configuration of shared O-RU. In multioperator case (or neural host mode), the O-RU calls the O-DU of tenant operator in hierarchical mode and the SMO of tenant operator in hybrid mode.

The O-RUs are shared by multiple O-Dus, so the O-DUs need to be able to oversee and supervise the O-RUs. Hence, a partitioned carrier occupied by an O-DU will not transmit any data when the O-DU's supervision is lost. The O-RAN alliance's solution ensures this, and the O-RU will cease transmission on the carrier when it experiences supervision failure with the associated O-DU. If the SMO has subscribed to receive alarm notifications, the O-RU will send an alarm to fault management function of SMO. This will then trigger alarm notification to non-RT RIC by the fault management function, which will be sent to the shared O-RU orchestration rApp.

The O-RU also needs to ensure management of its performance. Therefore, each O-DU will subscribe to the O-RU to receive performance management notifications regarding the operation of the fronthaul between the O-RU and the O-DU in concern. In this sub use case, the O-RUs will generate performance management report for each partitioned carrier and are sent to corresponding O-DUs. Note that it is necessary for the O-DU to first subscribe to receive such performance management notifications from the O-RU. The O-RU will send the notification periodically to the subscribed O-DU(s).

The O-RU needs to ensure resiliency of the system, for example, recovering from failures of various combinations of one or more O-DUs. This sub use case deals with the case where the O-DU that has configured the O-RU (i.e. active O-DU) is down, but the service needs to be maintained with other O-DUs (i.e. standby O-DU). The active O-DU fails, causing the shared O-RU to lose communication with this O-DU. The resiliency scenario depends on the plane that detects the loss of communication: M-plane or CU-plane. In the M-plane case, the O-RU first disables all the carriers associated with the O-DU m-plane connection, and then it issues alarms to its subscribers (can be other standby O-DUs). The alarm is forwarded to the managing entity e.g. SMO) that makes decisions on which standby O-DU will be the active O-DU to take over operation. Then, the new active O-DU performs configuration operation with the O-RU if necessary. In the CU-plane case, the difference is that the triggering of alarm notification is done on the CU plane, when the O-RU detects that it has lost CU logical plane or payload(s) with the active O-DU.

Finally, it is possible to operate and control ALDs in the shared O-RU use case. It should be clarified that the ALDs are not controlled by multiple O-DUs, but by one O-DU that operates the ALD controller on the O-RU. This is because there is only one ALD controller per O-RU, and consequently only one O-DU can be nominated as the ALD controller. This sub-use case starts with the shared O-RU orchestration rApp deciding which O-DU will be the ALD controller. The decision triggers non-RT RIC and configuration management function of SMO to configure the ALD of the shared O-RU. The SMO will then connect with the O-DU via O1 interface and configure the O-DU's ALD controller. The O-DU will configure the ALD-related aspects of the shared O-RU via the OpenFronthaul interface. During the operation of ALD, the O-DU will control the shared O-RU via the OpenFronthaul interface, and the O-RU will be interworking between the configuration over OpenFronthaul and the high-level data link control (HDLC) protocol that is used between the ALD and the O-RU.

6.3 Connected Mobility

6.3.1 Overview

The cellular communications can address many applications that require connectivity, but the applications that benefit most from the cellular communications are those that involve mobility of the user/thing. This is because the cellular communication technology was developed to address the mobility requirement of connected person, and it is makes most sense to consider the application of connected mobility when discussing how Open RAN can be applied. Therefore, this section will cover various use cases within the connected mobility application including vehicle-to-everything (V2X) use cases, unmanned aerial vehicle (UAV) use cases (also known as drones), and rail transport-related use cases.

6.3.2 Rationale

Before analyzing how Open RAN can be an enabler of the connected mobility, it is crucial to address the feasibility of the applications by identifying the market size. The market size of the use cases is important, because it describes the maximum potential revenue that can be generated from the use case. Although actual revenue generated per company may be different due to differences in target customers, it gives an idea of how promising the use case will be.

Looking at the V2X market, a forecast by ResearchAndMarkets.com indicates that it will reach up to US$ 1404.1 billion by 2030 with cumulative annual growth rate (CAGR) of 60.2% [10]. This means that the market size in 2022 is already as large as US$ 32.3 billion. While the figures include the market using other competing technologies (e.g. dedicated short-range communication), they portray the prospective potential of the V2X use case.

For the UAVs, the scope is confined to commercial drones, as consumer drones tend to be for light purpose use not requiring cellular connectivity. Commercial drone market forecast by ResearchAndMarkets.com indicates that the global market size will reach up to US$ 21.69 billion by 2030, over ninefold increase compared to the valuation of 2020 which was US$ 2.72 billion [11]. As in the case of the V2X market, this market size includes targets (the commercial drones) without cellular connectivity and can be seen as an overestimation. Nevertheless, the figures can be used to make a rough estimate of the potential of the UAV use case.

Finally, the potential of the rail-transport-related use cases can be fathomed by the smart railways market size. Although connectivity is just a component of smart railways, it is the fundamental enabler of smart railways and therefore the market size can be used as a compass to indicate the bright prospect of the use case. Again citing a market forecast by ResearchAndMarkets.com, the global market size for smart railways is expected to reach US$ 57.97 billion by 2030 [12], which is more than a double increase of the valuation of US$ 26.0 billion in 2021 [11].

6.3.3 Specific Use Cases

6.3.3.1 V2X

V2X is where the car's communication system is used to convey/receive information from external sources and internal sources (e.g. sensors) to assist driving and ultimately lead to autonomous driving. For example, the car will communicate with other cars, surrounding infrastructure (e.g. traffic lights), pedestrians' devices, and servers in data centers. V2X can be further categorized into six types of communication (see list and Figure 6.11)

- Vehicle to network or vehicle to cloud (V2N/V2C): the communication between a vehicle and the cloud (e.g. service servers), enabling various services and comfort functionalities in driving the vehicle. This mostly requires cellular connectivity.
- Vehicle to device (V2D): the communication between a vehicle and electronic devices connected to the vehicle, enabling interaction between driver's device such as the smartphone and the vehicle itself. This can be done without cellular connectivity.
- Vehicle to grid (V2G): the communication between a vehicle and the electricity grid infrastructure, enabling smart grid where battery of the car can be charged from the grid or discharged to the grid depending on the power demand on the grid. This can be done with or without cellular connectivity.

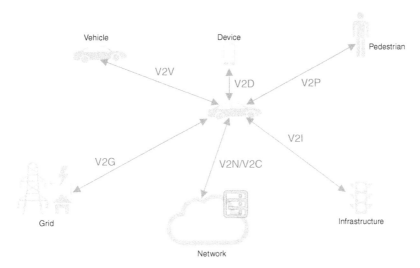

Figure 6.11 Six types of V2X communication. (Not all use cases and related depicted functions and terms are implemented in the current O-RAN specifications).

- Vehicle to pedestrian (V2P): the communication between a vehicle and the pedestrians, mostly in the form of handheld devices (e.g. smartphones). This enables applications such as detection and notification of collision. Cellular connectivity may be utilized in this type.
- Vehicle to infrastructure (V2I): the communication between a vehicle and the surrounding infrastructure, which can be the intelligent transport system overall, road signs, traffic lights, street lights, etc. This enables provision of real-time road information and lead to more optimized driving experience. Depending on the type of infrastructure, cellular connectivity is required.
- Vehicle to vehicle (V2V): the communication between a vehicle and other vehicles, enabling real-time exchange of context surrounding each vehicle and enhancement of safety performance. Cellular connectivity possesses features to support this scenario and therefore can be utilized.

The cellular communication can fulfill the requirements of V2X use cases overall. In the scenarios where the vehicle needs to connect to distant servers or infrastructure, cellular connectivity can fulfill the requirements. In some scenarios that involves short-range communications between a vehicle and other things around it, cellular connectivity may be used.

3GPP has already defined features to support various V2X use cases. There have been numerous enhancements to the cellular communications architecture, namely supporting V2X communications over PC5 (sidelink interface, where UEs can make direct communications between themselves) and Uu (radio interface, where the UE connects to the network) reference points. Further to the architectural enhancements, the 3GPP adds application layer support for V2X services, where the communication reference points above are used as transport to provide V2X capabilities. The V2X business relationship is broken down into the network (the mobile network operator), the V2X service provider, and the V2X user [13]. Accordingly, the enabler server can interact with various devices (not limited to vehicle, but also pedestrians and other types of infrastructure) directly or indirectly to offer different use cases.

O-RAN Alliance specifically highlights a scenario where Open RAN can facilitate V2X services. It is related to V2N/V2C use case, where the vehicle is traveling in high speeds and needs to interact with the cloud server or the network. Often, the vehicles are handed over frequently among different cells in these high-speed scenarios. This can cause suboptimal handover sequences, which can lead to anomalies such as short stay and ping-pong cases. This interferes with the experience and performance of the V2X applications, and the O-RAN Alliance utilizes AI/ML functionalities of near-RT RIC to customize handover sequence in the UE (i.e. vehicle) level based on past navigation and radio statistics data.

In this scenario, non-RT RIC, near-RT RIC, the RAN, and the V2X application servers are involved. The non-RT RIC is the entity responsible for building, deploying, and updating the AI/ML models into the near-RT RIC. It is also responsible for communicating intents, policies, and non-RAN data to the near-RT RIC. Near-RT RIC will, based on the inputs (AI/ML models and intents/policies) from non-RT RIC, achieve the desired performance/configuration. Near-RT RIC will also communicate configuration parameters to RAN and provide relevant data to the SMO for updating intents, policies, and AI/ML models. The RAN will receive optimization from the near-RT RIC to optimize the handover and provide required data to the near-RT RIC and SMO. V2X application server collects necessary data from the V2X UE and communicates the real-time traffic-related data about UE to non-RT RIC to enrich the analytics.

The high-level flow of handover optimization proceeds as follows. First, the SMO collects relevant data from the RAN and trains/updates the AI/ML model by the non-RT RIC. Then, the non-RT RIC will deploy the AI/ML model and traffic steering policy/intent into the near-RT RIC. The near-RT RIC will further collect data from the RAN nodes in order to make inference. The non-RT RIC also provides near-RT RIC with the enrichment data collected from the V2X application server. With these pretexts, the near-RT RIC will step in when anomalous handover has been detected, where relevant configuration and optimization will be sent to the RAN nodes to adopt. Relevant performance data will be collected by the non-RT RIC to update and train the AI/ML model further if required.

6.3.3.2 UAV

In a line-of-sight (LOS) manual control scenario commonly adopted in consumer drones, cellular communications is not a requirement. It is in the context of autonomous (or semi-autonomous) remote control of drones where 5G can provide key differentiation. For example, fleet of drones with centralized control can be used in a warehouse to automate inventory management [14] or fleet of autonomous drones can be used to scan a complex/large structure [15]. The capability of 5G to provide connection to multiple drones beyond LOS of a person (or many people) in charge can enable a new dimension of utilizing drones (Figure 6.12).

Specifically, low latency and the high bandwidth of 5G are critical in enabling these use cases. The low latency can really help leverage the advanced computing resources of the centralized controller (e.g. edge server) and overcome the limitations in computing resources due to lack of space in drones. High bandwidth is straightforward in that it enables

Extended coverage of control

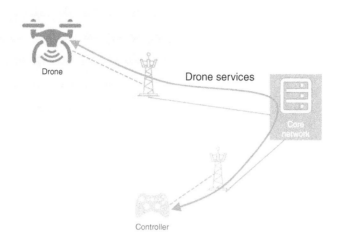

Figure 6.12 Commercial drones leveraging 5G. Source: Adapted from [16].

multiple drones to be connected to centralized controller or among themselves to coordinate their task.

Indeed, 3GPP has already defined 5G features to support connectivity for the drones (denoted UAV in 3GPP work items). It is possible to identify drones remotely and to authorize/authenticate drones to enable tracking and command and control of drones. It is also possible for drones to communicate with one another using the PC5 (Sidelink interface, is also used in V2X use cases) reference point. Similar to the V2X use cases, the drones also have application layer support that is defined to provide services for drones over 3GPP networks. More specifically, the enabler server can interact with different drones directly or indirectly to provide the application layer functionalities.

With focus on Open RAN, O-RAN Alliance has identified two key use cases related to drones: allocation of radio resource based on flight path and allocation of radio resources for scenario with drone control vehicle. The first use case deals with the optimization of radio performance for drones that are connected to cellular networks mostly optimized for terrestrial UEs. For example, uplink interference and limited propagation can be experienced. Therefore, the goal is to utilize the intelligence capability of Open RAN to avoid the above issues and provide adequate QoS for drones connected to 5G networks.

In this use case, non-RT RIC, near-RT RIC, RAN, and application server are involved. The non-RT RIC collects the necessary measurement data from the network and SMO to construct/train the AI/ML model and deploys the AI/ML model in near-RT RIC. It also sends policies/intents to near-RT RIC to ensure the model and optimization to not deviate into abnormal behavior. The near-RT RIC receives the AI/ML model and policies/intents from the non-RT RIC. The AI/ML models and policies/intents are executed in near-RT RIC, such that the radio resource allocation for the drones is optimized. The near-RT RIC also sends performance report to non-RT RIC for evaluation and optimization. The RAN nodes collect data on the UE performance (or drone performance in this case) and share it with the non-RT RIC via SMO. They also implement the optimization configuration/parameters. Finally, the application server provides enrichment information such as the flight path and climate information to be used by non-RT RIC in model training and forwarded to near-RT RIC for making inferences.

The high-level flow of this use case proceeds as follows. First, the non-RT RIC collects data from the RAN nodes via SMO and enrichment data from the application server. The non-RT RIC then trains AI/ML models, which are then deployed in the near-RT RIC. The non-RT RIC also makes further collection of enrichment data from the application server and updates the near-RT RIC with the enrichment data along with policies/intents. The near-RT RIC uses the data and the model to make inferences, leading to update of radio resource configuration by the RAN nodes. The RAN nodes may collect data and report to the non-RT RIC via SMO, where the data can be used to monitor/evaluate performance and update the AI/ML models.

The second use case, radio resource allocation for drone control vehicle scenario, deals with specific scenario as shown in Figure 6.13. Here, the drones have various devices (e.g.

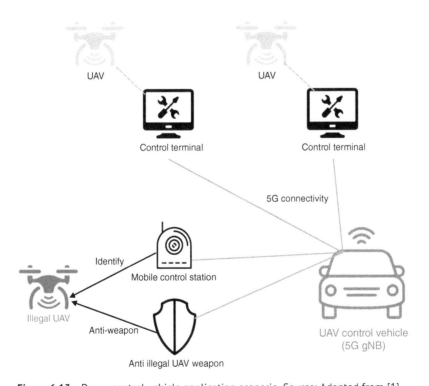

Figure 6.13 Drone control vehicle application scenario. Source: Adapted from [1].

cameras and sensors) mounted and are flying at low altitude and low speed. There are terminals for operation of these drones and there are anti-UAV systems that will identify and fight against illegal drones. The terminals and anti-UAV systems are connected to the drone control vehicle using 5G connectivity. This scenario can apply to specific applications such as forest inspection, pollution sampling, and HD live broadcast.

In this use case, non-RT RIC, near-RT RIC, RAN nodes, and application server are involved. Note that the drone control vehicle can contain all the functions above or may have all functions other than non-RT RIC co-located within the vehicle. The non-RT RIC receives UE-level radio resource adjustment requirements from the application server and sends resource allocation requirements and UE-level policies to near-RT RIC. The near-RT RIC receives the resource allocation requests and policies from the non-RT RIC, where it interprets and executes them accordingly. The near-RT RIC also communicates configuration parameters to the RAN nodes. The RAN nodes will receive resource allocation request from near-RT RIC, send registration information of terminals to application server and near-RT RIC, optimize radio resource allocation parameters, and adjust resource configuration parameters for different UEs. The application server receives registration information of terminals via SMO, collects user plane data from the RAN nodes, and sends UE-level radio resource adjustment requirements to non-RT RIC.

The high-level flow for this use case proceeds as follows. First, the non-RT RIC will request for allocation of radio resources, which is then processed by near-RT RIC and implemented by the RAN nodes to establish the cell. The RAN will notify the near-RT RIC and the application server (via non-RT RIC) of the registration information of the terminals. With this pretext, the non-RT RIC, based on the requirements of radio resource allocation adjustment from the application server, constructs resource adjustment policy. This is distributed to near-RT RIC and is converted to specific configuration parameter commands. These commands are sent to the RAN nodes and the RAN nodes will implement the changes in configuration parameters, where the uplink rate of a specific UE is adjusted.

6.3.3.3 Railway Communications

The railway community has already been using cellular communications as its communications system since the 2G era, where GSM-R technology, based on 2G GSM, was defined and adopted. In the era of digital transformation, the railway community is aiming to replace GSM-R with a successor technology, and 3GPP is already defining cellular communications-based solution to suit the needs of the railway industry. The requirements are defined by the Future Railway Mobile Communication System (FRMCS) functional working group of the International Union of Railways (UIC) [16].

The long adoption of cellular communications by the railway industry indicates that cellular communications is an appropriate technology to satisfy the industry's requirements. The reliability of cellular communications can ensure communications means for the rail workers that do not fail easily. Support for mobility allows continuity of the communications service even in mobile scenarios. The wide area network coverage can cover the massive magnitude of deployed rail infrastructure. The successor technology will enhance the capabilities even further and will also enable new use cases based on the use of multimedia and data communication.

Indeed, 3GPP has already translated many of the use cases and requirements defined by FRMCS. The following lists some of the use cases to illustrate how cellular communications can enable various applications in the railway industry [17, 18]:

- Voice communication between personnel: the traditional use case of cellular communication still serves as the basic block for the service. Relevant personnel can communicate with one another using FRMCS.
  o Train drivers and staffs on ground: the train driver may need to communicate with the staffs on ground (train stations or stationed in different depots), or vice versa.
  o Train drivers and controllers of the train: the train driver needs to communicate with the controllers of the train regularly.
- Emergency communication:
  o Railway internal: there are emergency alerts that need to be issued as per operational rules, and there can be cases where emergency communication among railway staffs are required.
  o Public emergency services: in cases where public emergency service is required (e.g. ambulance, firefighting, and police), the railway staffs need to be able to make emergency communication to these public emergency agencies.
- Operation of trains and infrastructure: maintaining the transportation networks and rail infrastructure is a challenge. Communications can assist the operation.
  o Coordination of multiple trains: with the location and speed data of trains that are received periodically, movement of trains can be coordinated to prevent accidents.
  o Trackside maintenance warning: use of mobile communications to inform the trackside maintenance staff on the approaching trains.
  o Remote control of trains (engines): the staff at ground can control the engine of the train if necessary.
  o Monitoring of train operation: train operation data related to performance and/or safety can be communicated to control centers for monitoring of operation.
  o Monitoring and control of infrastructure: communication is setup to monitor and control critical infrastructure, such as signals and indicators, barrier controls, movable infrastructure, and alarms.
- Smart stations: train stations are a major touchpoint to customers including passengers [18]. Providing smart services based on connectivity can enhance the satisfaction of the passengers and customers.
  o Allocation of platforms and management of trains: to increase track capacity, it is necessary to allocate the same platform to trains departing at different times. However, this comes with safety limitations where passenger boarding time and speed of incoming trains can limit the practice. Communication can be established between the trains and the station to optimize the flow.
  o Automatic monitoring of station: the data from closed circuit television (CCTV) and sensors can be inspected by AI to detect abnormal situation and issued warning to the station and the control office.
  o Smart kiosks: kiosk to provide information (e.g. ticketing, exits, train schedule, etc.) can be connected to other mobile devices (e.g. robots) to guide passengers in a more convenient manner.

- Next-generation services: next-generation features with enhanced capability of FRMCS system.
  - Real-time video communication: real-time video conferencing may be helpful in cases such as maintenance and reporting operation.
  - Transfer of CCTV data from train to ground: the CCTV data from the train needs to be transferred to the ground for recordkeeping. This is done when it is embarked at a station or at a depot.
  - Real-time automatic translation of languages: trains may cross borders, and it may be necessary to translate the communication between the control center and the driver for reducing ambiguity and enhancing safety.
  - Live streaming of media: media contents can be transmitted to the train from the ground for enjoyment.
  - Transfer of large media data from ground to train: there can be cases where large size multimedia databases (including movies, webpages, TV shows, etc.) need to be updated. This needs to be done when the train is stopped in a train station or a depot, where good connection is required.

Some of the use cases have already been selected by the SNCF of France, where SNCF is conducting preliminary studies on rolling the technologies out on the field. According to the SNCF [19]:

1. trains need fast, reliable connections to deliver critical services and meet system requirements for safe operations, such as command and control signaling, driverless trains, and more;
2. passengers need broadband connectivity to go online during their journey;
3. the many sensors installed in trains and rail infrastructure make maintenance easier, and the maintenance work in our technicenters relies increasingly on connected objects;
4. stations are an integral part of the "smart city" concept powered by 5G;
5. 5G will bring connectivity to SNCF facilities in areas with poor coverage.

As for the solution to realize the requirements, the Mission Critical Push-To-Talk (MCPTT) architecture is enhanced to support most of the use cases of FRMCS. Some of the use cases require location. Note that the 3GPP work is still ongoing as of May 2023, with a related study item is planned to be completed by September 2023, and migration/interconnection aspects important in implementation are to be completed by December 2023.

In railway communications, the Open RAN can contribute by leveraging its key differentiators. First, the intelligence capability can assist in the mission critical scenarios. Radio performance can be optimized by AI/ML models in near real-time manner. In addition, the intelligence can help in handover scenarios as was the case in V2X, enhancing the QoS for the applications. Second, opening up of interfaces and disaggregation of layers can add scalability and flexibility of the FRMCS system overall. Rather than taking a long time to install and operate increased capacity, the system's capacity can be increased by adding hardware components or installing new software depending on what is necessary.

6.4 Private Networks

6.4.1 Overview

Private network is when a network is built for the organization's own private use. For example, it maybe that an airport wants to deploy its own network such that its staffs can communicate in a mission-critical manner, and its passengers can enjoy data connectivity. Or it maybe that a factory wants to establish a network that will solely be used by its staffs or its machines. These are all examples of private network.

The concept of private network is not new. There were already cases of a network dedicated for an organization even back in the days of fixed circuit-switched phones. There was private branch exchange (PBX) that could handle calls within the organization and that can be used by the organization's authorized users only. Same holds for mobile networks and the concept became even more popular in the days of 4G LTE (as of February 2022, 61% of private networks deployed in the world were based on LTE only [20]). In 5G, the 3GPP specifications explicitly support private networks and even those that combine privately deployed network and public mobile networks.

As the industry expects Open RAN to be a great enabler for the private networks, it is essential to consider the private networks in detail and how Open RAN can assist in deployment and operation of private networks. In this context, this section covers the definition and the rationale of private networks, the role of Open RAN in enabling private networks, and selected application of private networks.

6.4.2 Introduction to Private Network

6.4.2.1 Definition

Private network is defined to be an isolated network deployment that does not interact with a public network according to the 3GPP [21]. However, this does not give a full picture of the various types of private networks available today, because it only focuses on the deployments that are completely private. In both reality and standards, there are deployments that utilize aspects of the public mobile network.

To account for this variety, the 3GPP uses terminology of non-public network (NPN) in its specifications. The NPN is a network that is intended for non-public use [21] and can be either fully deployed privately or deployed utilizing some parts of the public mobile network. The former is referred to as stand-alone non-public network (SNPN), not relying on network functions provided by a public mobile network, and the latter as public network integrated NPN (PNI-NPN), deployed with the support of public mobile network. The relationship among the public network, PNI-NPN, and SNPN is as depicted in Figure 6.14. The blue color indicates NPN, and the two variations of green color distinguish the two types, while public network is fully distinguished from NPNs.

The choice between SNPN and PNI-NPN depends on the requirement of the organization. In SNPN, the organization will be able to have more control on the network (e.g. full control over the devices and subscriptions, maintenance of networks, and security) but will incur more initial investment and as well as requiring own/unlicensed spectrum. On the other hand, PNI-NPN relies on the public mobile network for many aspects so the control

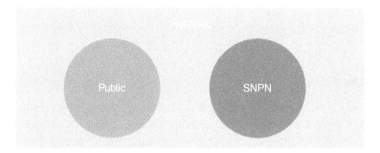

Figure 6.14 Difference between public networks and two types of NPN.

will be limited. However, it will come at a lower initial investment and less complexity in maintenance and services. In any case, both types require some sort of special treatment that a subscriber of public mobile network will not be able to benefit from, hence justifying the need for deploying and operating NPN.

Before progressing further, it should be clarified that the definitions above were covered to distinguish different types of private networks. In the subsequent sections and chapters of this book, the term private network will be used to indicate what was referred to as NPN above. PNI-NPN will be used if there is a need to distinguish SNPN from PNI-NPN, but most of the discussions in the book will stick with the term private network.

6.4.2.2 Rationale

The benefits of private networks can be classified into two categories. The first category of benefits relates to increase in performance and the second related to increase of control over the network. For the former, the private networks can provide improved coverage, as the network (or part of it) is deployed to suit the needs of the organization and not the public mobile network operator. The private network is more likely to provide lower latency, because many of the network functions and equipment are likely to reside in the premise of the organization and therefore physical latency is reduced. In addition, private network is more likely to be more reliable, as in the case of lower latency. This is because there are less points of failure outside of the organization's control. There is also a potential to reduce cost with private networks, because while there is more initial investment, the operating cost of the network can be less than renting a network.

The second category of benefits include increased security, as the traffic received differentiated treatment (in some cases, they remain in the organization's premise). In the same context, data privacy can be achieved. The private networks can also allow the organizations to formulate and apply policy in traffic prioritization and congestion management. That is, the organizations can choose which of its traffic to prioritize and how traffic should be managed, allowing more refined control of the network and hence have the network adapted to the organization's purpose.

These benefits explain the increasing market size of private networks. ResearchAndMarkets.com forecasts that the global private 5G network market will reach US\$ 36.08 billion by 2030 [22]. Considering the CAGR of 47.5%, the 2022 market size is estimated to be US\$ 1.61 billion. Analysys Mason forecasts the global private network (not limited to 5G) market size will grow to US\$ 7.7 billion by 2027, where there will be up to 39,000 private networks deployed around the world [23].

6.4.3   Role of Open RAN

While private network per se is a promising use case, it is also a worthwhile topic to be discussed in the context of Open RAN. Many in the industry expects Open RAN to be the enabler for private networks due to the following reasons: open and flexible architecture, open ecosystem for developers, great potential to be integrated with edge cloud services, and ability to address grade of quality that private networks demand.

Open and flexible architecture is the core concept of Open RAN, but it also benefits the private network deployments by making the technology more accessible to the enterprise. Unlike the long-established use cases and requirements of public mobile networks, the enterprises have diverse requirements and most of times less complex use case. In this context, Open RAN can make the private networks more easily deployable and manageable with the open reference designs that are more similar to enterprise systems (e.g. Wi-Fi) than traditional cellular communication is [24].

The open interfaces, together with the open and flexible architecture, foster an open ecosystem for developers. With the flexibility and more IT-like technology, it can be easier for developers to become familiar with cellular technology and develop new solutions [25]. The solutions will not be monolithic, but will most likely to cater to multiple segments, where target segment's willingness to pay, physical environment, and use cases will differ [24]. This leads to increased value of the mobile private network for the organization, forming a virtuous cycle of growth in private networks and solutions for private networks.

The Open RAN is also seen as a technology that can integrate with edge cloud services more easily than the traditional RAN can [26]. This stems from its open architecture and open interfaces, where it is easier to program the RAN and therefore is easier to be integrated with edge cloud services. Since private networks are likely to utilize edge cloud services that are more suitable for their requirements than the public cloud services are, this characteristic makes the Open RAN an attractive candidate for private networks to consider.

Enterprises and other entities interested in deployment and operation of private networks benefit from techno-economic assessment of the variants that comply with their requirements. It is thus important to understand the pros and cons of each model, as described in [27]. Open RAN, as it matures, may offer an attractive alternative to the more traditional models of private network realizations.

Most importantly, although networks are critical to the enterprise's business, the performance required by the private networks is not as stringent as the demand on the public networks. For example, the public network may need to support high bandwidth communications to thousands of people crowded in a city, but the private network is likely to not be faced with such challenges. Furthermore, the high performance of traditional technologies is sometimes based on the assumption of public network's carrier-grade requirements and use cases (e.g. roaming and charging), but some private networks may not require those requirements and use cases. In this context, the solution that will cater to the private networks will be a lower cost technology that fulfills less of the public network requirements but address the organization's pain points.

6.4.4 Applications

6.4.4.1 Smart City

In discussing the future of cities, many refer to the smart city to be the future. According to European Commission, smart city is a place where traditional networks and services are made more efficient with the use of digital solutions for the benefit of its inhabitants and business. The smart city therefore requires connection of various things and people to the network and digital solutions that process the requests and the data from the things and people. The construction of smart city is not a big issue when the city is built from the scratch, as the ICT infrastructure can be considered in urban planning and construction of buildings. For the established cities, however, it is difficult to deploy connectivity infrastructure as civil works can be limited due to old buildings and old infrastructure. Furthermore, the need to maintain landscape for esthetics and citizens' welfare may need to be sacrificed when installing new connectivity equipment. This makes realization of smart cities a very complicated project in real life.

Indeed, the city of Dublin has put forth its requirements for connected city in its initiative with Telecom Infra Project [28]. They explicitly highlight that, despite the strong interest in providing high-quality mobile network services for better quality of life and business satisfaction, the high-quality environment and an orderly planning process are required to avoid undesirable consequences harming the city landscape and city life. For example, the street assets may be filled with equipment or disruptive civil works can occur concurrently due to uncoordinated installation. While the requirement document focuses on the coordination of stakeholders and assets, there is an important technical consideration that arises from the city's needs. That is, it can be more sightly and efficient to have a network equipment installed in a site where the space is limited (e.g. street assets) and have different network operators connect to the equipment and share it to provide their service. The Open RAN is not only about opening up architecture and interfaces but also about being able to share the network among different entities. In this respect, the Open RAN can address the city's needs to strike the balance between good connectivity and urban planning. In addition to this, the smart city infrastructure needs to be well integrated with different types of IT services, ranging from public cloud-based services to edge cloud-based services. With the open and flexible architecture, the Open RAN can be an attractive technology candidate for providing connectivity of smart cities.

6.4.4.2 Industry

Private networks for industrial purposes can also be a promising use case for Open RAN. Here, the industrial purposes are loosely defined as manufacturing, where the cellular connectivity can be applied to manufacturing processes and unlock efficiency/yield gains in conjunction with IT services including AI. While there can be various types of manufacturing, the common ground of the user organizations is that cellular connectivity is just a means to enhance their quality and efficiency, and the connectivity is not their core competence. This means that they may not necessarily be proficient in ICT. Furthermore, the production sites tend to be local and the size of the network to be deployed can be small.

In this context, the Open RAN can provide an attractive technology solution for these companies. First, the ability for Open RAN to be integrated with edge cloud services provides incentives for these companies to adopt. As mentioned earlier, they are not likely to be proficient in ICT, and having the ability to be integrated with edge cloud services can potentially reduce hassle in integration and management of ICT solutions. Second, the intelligence of Open RAN can provide near-real-time optimization of the connectivity according to the manufacturing use cases and scenarios the companies require. Third, the network requirement for these companies will be leaner and more different from that of the public network operators. In addition, the scale of deployment will be small for these companies, making the business unattractive for the traditional incumbent vendors with traditional technologies. The Open RAN technology and the Open RAN vendor can tailor the requirements to non-carrier-grade requirement to provide more customized solution. Furthermore, the Open RAN vendors are usually new entrants and therefore small-scale markets may still be profitable for them.

6.5 Potential for the Future

6.5.1 Key Differentiators of Open RAN Revisited

The Open RAN use cases are not limited to the ones that were covered in the previous sections. Any use cases that can benefit from the differentiators of Open RAN are prospective use cases of Open RAN. It will be therefore beneficial to revisit the key differentiation provided by Open RAN. First, open interfaces enable the mix and match of RAN equipment and multi-vendor configuration. The open interfaces also allow more developers and users to access the network technology and to develop/tailor solutions for networks. Second, virtualization enables RAN to disaggregate software and hardware layers, thereby reducing cost, ensuring scalability, and allowing better integration with edge cloud services. Finally, intelligence of Open RAN enables automated management of the RAN based on big data analytics and AI technologies. This is a crucial differentiator as it allows optimization of the RAN with least intervention from humans.

6.5.2 Potential Use Cases

It is not an exaggeration to emphasize that there will be more use cases of Open RAN than the ones stated in the previous sections and the ones that will be stated in this section. Nevertheless, this section presents a list of potential use cases, selected from 3GPP's on-going work in Release 18 and Release 19 studies that may benefit from Open RAN.

- Localized mobile metaverse services [29]: this refers to the use of metaverse (persistent, shared, perceived set of interactive perceived spaces) in the context of particular location but able to be accessed anywhere. Examples include virtual sports, government building providing augmented reality services to improve public access, and avatar services. This requires support of interactive AR/VR media shared among multiple users in a single location and to use/expose local information to third parties to enable metaverse services. The Open RAN's intelligence capabilities can optimize the RAN performance to suit metaverse services, and open interfaces can facilitate exposure of RAN to enable metaverse services.

- Support for XR services: this refers to the use of XR technologies to provide media services over 5G. Since such media traffic has different characteristics (e.g. some frames of image are more important than others) from traditional data communications, architectural enhancements to the 5G system in core network and RAN are necessary [30]. In the context of RAN, the RAN needs to be able to apply the policy given by the core network and handle uplink packet data accordingly. The RAN also needs to coordinate uplink–downlink transmission to meet round-trip latency requirements. The Open RAN's intelligence capabilities can enforce the core network policies effectively by intent/policy mechanisms and by applying AI/ML models to optimize based on the available usage data.
- Vehicle-mounted relays [31]: this refers to the 5G relays that will extend the coverage of 5G networks that are mounted on vehicles, making the relay mobile and/or periodic depending on the type of vehicles. The relays can be mounted on dedicated vehicles in emergency situations or in crowded situations such as sports events. They can also be mounted on the public transports (buses or trains). While the intelligence capabilities of Open RAN can surely help in the temporary situations such as emergency and sports events, the potential can be exploited in the case of public transports. This is because the public transports travel in periodic and regular cycles, and it is very likely for users around and in the public transport to exhibit similar usage patterns. The AI/ML models can be trained and updated to optimize the radio performance significantly.

References

1 O-RAN Alliance, "O-RAN.WG1.Use-Cases-Detailed-Specification-v09.00," 2022.
2 Connectivity Technology Blog, "CSI-RS vs SRS Beamforming," 02 03 2021. [Online]. Available: https://www.connectivity.technology/2021/02/csi-rs-vs-srs-beamforming.html. [Accessed 16 04 2023].
3 Ibrahim_Sayed, "Does massive MIMO Beamforming Optimization differ in Open RAN?," 07 2022. [Online]. Available: https://www.telecomhall.net/t/does-massive-mimo-beamforming-optimization-differ-in-open-ran/18721. [Accessed 16 04 2023].
4 5GWorldPro.com, "What is multi-user MIMO (MU-MIMO) in 5G?," 28 05 2020. [Online]. Available: https://www.5gworldpro.com/5g-knowledge/what-is-multi-user-mimo-mu-mimo-in-5g.html. [Accessed 16 04 2023].
5 5GWorldPro.com, "What is the Dynamic spectrum sharing," 22 02 2020. [Online]. Available: https://www.5gworldpro.com/blog/2020/02/22/183-what-is-the-dynamic-spectrum-sharing/. [Accessed 16 04 2023].
6 F. Granelli, "Chapter 3 - Network slicing," in *Computing in Communication Networks: From theory to practice*, Academic Press, 2020, pp. 63-76.
7 Telecom Infra Project, "Network as a Service (NaaS) Site Economics," 2021.

8 GSMA, "GSMA Infrastructure Sharing: An Overview - Future Networks," 18 6 2019. [Online]. Available: https://www.gsma.com/futurenetworks/wiki/infrastructure-sharing-an-overview/. [Accessed 16 03 2023].

9 K. Murphy, "Centralized RAN and Fronthaul," Ericsson, [Online]. Available: https://www.isemag.com/wp-content/uploads/2016/01/C-RAN_and_Fronthaul_White_Paper.pdf. [Accessed 6 July 2019].

10 Business Wire, "Global Vehicle-to-Everything (V2X) Market 2022-2030: Sector to Reach $1.4 Trillion by 2030 at a 60% CAGR - ResearchAndMarkets.com," 06 04 2023. [Online]. Available: https://www.businesswire.com/news/home/20230406005354/en/Global-Vehicle-to-Everything-V2X-Market-2022-2030-Sector-to-Reach-1.4-Trillion-by-2030-at-a-60-CAGR---ResearchAndMarkets.com. [Accessed 12 04 2023].

11 Business Wire, "Global Commercial Drones Market Opportunity Analysis and Industry Forecasts, 2021-2022 & 2030 - ResearchAndMarkets.com," 25 05 2022. [Online]. Available: https://www.businesswire.com/news/home/20220525005498/en/Global-Commercial-Drones-Market-Opportunity-Analysis-and-Industry-Forecasts-2021-2022-2030---ResearchAndMarkets.com. [Accessed 12 04 2023].

12 Business Wire, "Global Smart Railways Market Report to 2030 - by Type, Component, and Region - ResearchAndMarkets.com," 20 07 2022. [Online]. Available: https://www.businesswire.com/news/home/20220720005521/en/Global-Smart-Railways-Market-Report-to-2030---by-Type-Component-and-Region---ResearchAndMarkets.com. [Accessed 12 04 2023].

13 3GPP, "TS 23.286 Application layer support for Vehicle-to-Everything (V2X) services; Functional architecture and information flows v17.4.0," 2022.

14 GSMA 5G Transformation Hub, "How 5G can enable drones to perform key tasks," 2022.

15 GSMA 5G Transformation Hub, "How 5G drones will help deliver digital twins," 2022.

16 3GPP, *SP-190064 New WID on Application Architecture for the Mobile Communication System for Railways Phase 2,* 2019.

17 3GPP, "TR 22.989 Study on Future Railway Mobile Communication System v19.2.0," 2022.

18 3GPP, "TR 22.890 Study on supporting railway smart station services v19.0.0," 2022.

19 SNCF, "Making the Rail System Better—with Telecoms," 03 01 2023. [Online]. Available: https://www.sncf.com/en/innovation-development/innovation-research/digitalization/telecoms-rail. [Accessed 16 04 2023].

20 GSA, "Private Mobile Networks," 2022.

21 3GPP, "TS 22.261 Service requirements for the 5G system v18.6.1," 2022.

22 Business Wire, "Global Private 5G Network Market Analysis Report 2022: A $36 Billion Market by 2030 - Robust Deployment of Next-Generation Private 5G Network Infrastructure Across Smart Cities - ResearchAndMarkets.com," Business Wire, 01 06 2022. [Online]. Available: https://www.businesswire.com/news/home/20220601005762/en/Global-Private-5G-Network-Market-Analysis-Report-2022-A-36-Billion-Market-by-2030---Robust-Deployment-of-Next-Generation-Private-5G-Network-Infrastructure-Across-Smart-Cities---ResearchAndMarkets.co. [Accessed 14 04 2023].

23 Analysys Mason, "Spending on private networks will reach USD7.7 billion in 2027, but challenges to adoption persist," Analysys Mason, 17 10 2022. [Online]. Available:

https://www.analysysmason.com/research/content/articles/private-networks-forecast-rma17/. [Accessed 14 04 2023].

24 Analysys Mason, "Open RAN vendors have an opportunity with private networks, regardless of their success with public 5G," 2022.

25 Mavenir, "Private 5G: A new paradigm for network operators," 2021.

26 G. Brown, "Open RAN for private 5G and venue networks," LightReading, 19 01 2022. [Online]. Available: https://www.lightreading.com/open-ran-for-private-5g-and-venue-networks-/a/d-id/774702. [Accessed 13 04 2023].

27 Jyrki Penttinen, Jari Collin, Jarkko Pellikka, "On Techno-Economic Optimization of Non-Public Networks for Industrial 5G Applications," IARIA AICT 2023, 26 June 2023. [Online]. Available: https://www.thinkmind.org/index.php?view=article&articleid=aict_2023_1_10_10003. [Accessed 20 September 2023].

28 Telecom Infra Project, "Connected City Infrastructure Requirements for Dublin: TIP Requirements Document," 2021.

29 3GPP, *SP-220353 New SID on Localized Mobile Metaverse Services (FS_Metaverse)*, 2022.

30 3GPP, *SP-221326 New WID: Architecture Enhancements for XR and media services (XRM)*, 2022.

31 3GPP, *SP-230105 Updated WID on Architecture Enhancements for Vehicle Mounted Relays (VMR)*, 2023.

7

Open RAN Security Aspects

7.1 General

Open RAN changes the "traditional" ways the mobile communications networks have been designed and deployed. The disaggregation of the new concept brings many benefits through the more open environment, such as the possibility to include more stakeholders of variety of domains. On the other hand, the security of the Open RAN needs to be at least equally good, or preferably better, than on any of the previous generations [1, 2].

The openness of the new network model is done through virtualization of the network functions, and the deployments are based on cloud infrastructure and operations. This provides means to disaggregate software from the hardware it runs upon as instances [3–5].

So, Open RAN standards disaggregate the radio access network with the idea of changing the proprietary protocols and interfaces into more commonly standardized, and thus better interoperable form. This controlled disaggregation is the key for ensuring diversity of vendors that can offer their products as "mix and match" – the aspect that was almost unheard in practice before.

Open RAN builds upon the security of 5G and provides operators tools to control and enhance the network security. The mobile network operator has access to the complete network performance information through the disaggregated components that connect via open interfaces, revealing network activities that have in certain extend been previously internal for merely vendors' information. One of the benefits of this aspect is the visibility to security events, and thus opportunity to react to potential security threats as early as possible. Combining more easily available information on the state of network functions through open interfaces and related security events data helps operators analyze more rapidly the root cause of such events [6].

The security of Open RAN concept involves many aspects of which this chapter summarizes some of the most concrete and important ones.

7.2 User Equipment

7.2.1 SIM

Subscription identity module (SIM) has been used since the deployment of 2G GSM (Global System for Mobile Communications). Along with newer mobile communications

Open RAN Explained: The New Era of Radio Networks, First Edition.
Jyrki T. J. Penttinen, Michele Zarri, and Dongwook Kim.
© 2024 John Wiley & Sons Ltd. Published 2024 by John Wiley & Sons Ltd.

generations, the basic principle has not changed. It houses the unique identifiers related to subscription, such as the subscription key, that are also stored in the secure repository of the home network. The critical information such as the subscription key does never leave the SIM nor the safe repository, but instead, further keys are derived for the communication sessions.

Prior to the establishment of the data, voice, messaging, or signaling channel, the mobile network authenticates and authorizes the subscriber to use the services. The development of ever-increasing attack methods, the procedures for this, as well as for the actual encryption methods for the communications and signaling channels, have been designed to cope with the changing security and trust landscape. As an example, mutual authentication for subscription and network was adopted based on the practical experiences of man-in-the-middle attacks that were possible to carry out in less protected environments so that bad actors could deploy fraudulent base stations pretending to be legitimate network components. Also, the key lengths and other aspects have been evolved along with more powerful tools, fraudsters have access to, and cryptographical methods include today also integrity protection to prevent fraudulent injection of false messages to the network.

Also, physical SIM has been upgraded along the years to provide increased security such as protection mechanisms against side channel attack. SIM works still well as a hardware-based secure element, providing typically much better platform compared to software-based methods.

SIM has also evolution path as for the physical aspects. The form factor of SIM has resulted in smaller plugin cards and embedded variants that device vendors can solder into customer devices as a separate safety component. The latest development also provides a completely integrated SIM variant that resides in some of the existing components of the device such as central processor. The management of the permanently embedded and integrated eSIM can be done by remote SIM provisioning (RSP) concept that, e.g. GSM Association (GSMA) has aligned providing technical and procedural specifications to the ecosystem for interoperable functioning.

As for Open RAN, the subscription management and protection mechanisms based on traditional plugin SIM or embedded or integrated variant, eSIM, can be considered agnostic to the network functionality. Nevertheless, the basic challenges with the security attacks remain valid so it is important for the Open RAN ecosystem to ensure the infrastructure does not open new attack vectors compared to "traditional" network models.

7.2.2 Device

Open RAN per se is transparent for the functions of the user equipment (UE), so by default, they continue working in Open RAN environment without need for modifications.

7.3 Current Security Landscape

7.3.1 Overview

This section lists relevant entities and resources dealing with the security aspects of Open RAN. This section also discusses physical radio interference attacks and protection and presents case studies and shielding recommendations.

As stated in [7], operators cooperate with national authorities to share information on security, implementation, and management of Open RAN. As the landscape evolves, operators have also requested to include Open RAN to be part of the GSMA Security Assurance Scheme (NESAS) as well as the European Union Agency for Cybersecurity (ENISA) certification scheme of the European Union (EU). Reference [7] further indicates that as an example, Deutsche Telekom, Orange, TIM, Telefónica, and Vodafone are committed to apply mandatory controls defined by the O-RAN Alliance and 3GPP security specifications including the supply chain.

The key in this environment would be for the ecosystem to apply a zero-trust approach covering all the respective vendors fortified by the standards and specifications of 3GPP and O-RAN Alliance. It is also important to include the national authorities' requirements to adequately assess risk profiles of vendors.

The Telecom Infra Project (TIP) complements the landscape for the disaggregated RAN components. Of the resources of TIP, the base station requirements outline architectures for the RU, DU, and CU, as an example, of the use case for small cell 5G NR supporting NSA and SA as an underlay to the existing macro cell architecture providing long-term evolution (LTE) and NR services [8–10] TIP has produced also many more documents on disaggregated mobile networking, including trial experiences and Open RAN productization aspects. Nevertheless, these resources do not focus on security aspects but refer to Open RAN Alliance and other entities.

In this chapter, select government entities are referred. To complement the filling of potential gaps in the security landscape, O-RAN Alliance's Security Work Group (WG11) has published security control mechanisms for select interfaces in March 2023 that are discussed in this chapter, too.

7.3.2 Open RAN Work on Security

7.3.2.1 Context and O-RAN Specifications

The security of Open RAN is of utmost importance in global level and cross ecosystem stakeholders. The principle of the adequate security level should apply to Open RAN systems already before commercial roll-out starts taking place. The O-RAN Alliance Working Group 11 (WG11), which was Security Focus Group (SFG) previously, defines the security architecture with the aim to facilitate the applicability of common security standards for O-RAN systems.

The practical deployment scenarios need to consider the potential threats. As stated in [11], the portfolio of radio units and frequency bands can differ between networks, but the 4G and 5G design considerations will remain equally applicable. Furthermore, site integration aspects are the very key to secure seamless coexistence with possible legacy 2G and/or 3G network equipment located at the site, including the security controls to combat against the threats to older technologies.

As stated by O-RAN Alliance [12], the attack surface of RAN systems is expanded as a result of the use of open- and cloud-based architectures. At the same time, thanks to the transparency of the applied open interfaces, it will be easier to monitor potential vulnerabilities and failures.

This open environment is beneficial also in a form of increased competition, as various O-RAN equipment and solution vendors will be able to provide and implement their hardened security so that the operators will be able to evaluate and select the most

Table 7.1 O-RAN security-related specifications and complementing documents.

O-RAN document	Version, date
O-RAN Security Requirements	V5.0, March 2023
O-RAN Security Protocols Specification	V5.0, March 2023
O-RAN Security Threat Modeling and Remediation Analysis	V5.0, March 2023
O-RAN Study on Security for O-Cloud	V2.0, March 2023
O-RAN Study on Security for Application Lifecycle Management	V1.0, March 2023
O-RAN Study on Security Log Management	V1.0, March 2023
O-RAN Study on Security for Service Management and Orchestration (SMO)	V1.0, March 2023
O-RAN Study on Security for Shared O-RU (SharedORU)	V1.0, March 2023
O-RAN Study on Security for Near-Real-Time RIC and xApps	V2.0, March 2023
O-RAN Security Test Specifications	V3.0, October 2022
O-RAN Study on Security for Non-RT RIC	V1.0, July 2022

adequate ones complying with their requirements. The EU 5G Toolbox brings further aid to consider the security recommendations.

The O-RAN Alliance SFG considered Open RAN system risks are based on the ISO 27005 risk analysis methodology and applying a Zero-Trust Architecture as per the National Institute of Standards and Technology (NIST) principles [13]. O-RAN Alliance released the initial Security Test Specifications in March 2022.

As stated in [13], the SFG work was materialized in four security specifications. These serve as the building blocks to the O-RAN security architecture. The initial SFG specifications were from 2021, accessible on the O-RAN Alliance website at https://www.o-ran.org/specifications [14].

The O-RAN Alliance WG11 documentation includes the ones summarized in Table 7.1.

The following sections summarize the key contents and focus of the O-RAN security specifications.

7.3.2.2 O-RAN Security Requirements

The security requirements document outlines requirements and controls for O-RAN interfaces and network functions [12]. The focus of the document is on security requirements and security controls for O-RAN interfaces and network function.

The document includes security requirements for O-RAN network functions and applications, O-RAN interfaces, and other transversal requirements. The latest version 5.0 includes security control mapping for RIC application programming interfaces (APIs), security requirements for SMO logging, and O-Cloud security requirements such as image security requirements for Apps, VNF, and CNF (cloud-native network function), and image protection. The O-Cloud requirements also include secure storage, chain of trust, software image protection, and controls for virtualization and isolation.

The document also describes guidelines for generation, delivery, and use of a software bill of materials (SBOM) for O-RAN. Reference [12] notes, the U.S. Department of Commerce

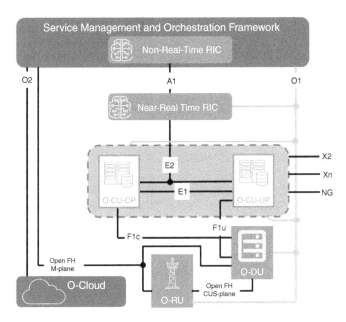

Figure 7.1 O-RAN logical architecture. Source: Adapted from O-RAN Alliance [5].

(DoC) and the National Telecommunications and Information Administration (NTIA) define SBOM as a formal record containing the details and supply chain relationships of various components used in building software. SBOM is thus important component in well-defined software development lifecycle (SDLC) process and serves as an industry best practice of secure software development. It considers also the software supply chain and ensures handling of safe handling of vulnerability notifications and updates across the customer base. The SBOM is maintained and used by the software supplier and stored and viewed by the network operator.

Finally, the informative Annexes of the document outline the security principles mapping to security requirements, a list of 3GPP security requirements, and guidance on security requirements with potential solutions.

Figure 7.1 summarizes the O-RAN logical architecture and the interfaces the O-RAN security requirements document focuses on.

7.3.2.3 O-RAN Security Protocols Specification

The security protocols specifications document defines the security protocols for O-RAN compliant implementation. These are secure shell (SSH), transport layer security (TLS), NETCONF over secure transport, IPSec, CMPv2, and OAuth 2.0. The document also presents the cryptographic operations for O-Cloud Network Functions and Apps package protection.

The SSH is used in the O-RAN and 3GPP interfaces for authentication, confidentiality, and integrity. The document provides instructions for the enabling and disabling cipher types and to configure backward compatibility.

For the TLS implementation, the document instructs the use of TLS 1.2, and support of TLS 1.3.

The NETCONF over secure transport is used by the service management service providers and consumers.

The datagram transport layer security (DTLS) is implemented in O-RAN and 3GPP interfaces for mutual authentication, integrity protection, replay protection, and confidentiality protection based on DTLS 1.2. It is specified that the DTLS 1.0 shall not be used.

IPSec can also be used for the authentication, confidentiality, and integrity in interworking and appliance rules. The document provides instructions to minimize the options to better cope with the interoperability.

CMPv2 (Certificate Management Protocol version 2) is based on Internet X.509 Public Key Infrastructure (PKI) Certificate Management Protocol (CMP) as per IETF RFC 4210 and 4211. It is used for the connection between network element (NE) and the CA of an operator as described in the 3GPP TS 33.310.

OAuth 2.0 is authorization framework that the IETF RFC 6749 describes. It is meant for the authorization of the requests of service consumers to the service producers.

7.3.2.4   O-RAN Security Threat Modeling and Remediation Analysis

O-RAN architecture differs from the previous 3GPP models considerably with its new components and interfaces, open-source code, virtualization, and increased inclusion of artificial intelligence (AI) and machine learning (ML). The new concept allows increased set of stakeholders to form part of the ecosystem and allows advanced business models. This openness and expanded landscape means that there also will be new threats.

The O-RAN security threat modeling and remediation analysis describe this evolved environment and the items that the involved parties need to identify and consider, such as vulnerabilities, threats, requirements. recommendations, and countermeasures, with the aim to present sufficiently concrete view on the organization of security in O-RAN, assuming the 3GPP security requirements are met.

The document references best practices by various entities, such as 3GPP, ETSI, NIST, ENISA, and GSMA. The contents include the description of roles of stakeholders involved in O-RAN and presents prerequisites for the secure implementation and execution of O-RAN system. The document continues addressing threat model by identifying threat agents, surfaces, vulnerabilities, and component and interface-specific threats. It also discusses security principles for countermeasures and finally outlines threat risk assessment principles.

7.3.2.5   O-RAN Study on Security for O-Cloud

Study on security for O-Cloud describes the O-RAN cloud architecture with its components, interfaces, and critical services and outlines cloud deployment scenarios such as infrastructure, platform, and software as a service models, referred to as IaaS, PaaS, and SaaS, respectively.

IaaS enables capability for operator to provision computing resources such as processing, storage, and network. In this environment, operators and vendors can deploy and execute software including host OS, virtualization platforms, and operator-specific virtual network functions (VNFs) and CNFs.

PaaS enables operators' capability to deploy VNFs and CNFs on cloud by programming languages, libraries, services, and tools that the cloud provider supports. PaaS allows operators to develop and execute VNFs and CNFs so that they do not need to construct or maintain a separate platform.

SaaS is the cloud consumers' capability to use cloud providers' applications that run on cloud so that applications can be accessed through web browser or programming interface. As the concept is based on shared resources, some of the potential security risks of the SaaS model include limited visibility for logging at low layers, lack of transparency for data security, low control, and insider fraud. Nevertheless, the model can be applicable in, e.g. private network environment so that the cloud consumer, served by cloud provider, assumes the responsibility to secure data and access to applications.

The deployment models cover private, community, public, and hybrid cloud environments. The document also discusses the roles and responsibilities, security problem definition, and recommendations and best practices. Finally, the document outlines the means for risk assessment.

Figure 7.2 summarizes the cloud architecture.

As presented in Figure 7.2, the following components form part of the Cloud:

- SMO components that are Federated O-Cloud Orchestration and Management (FOCOM), Network Function Orchestrator (NFO), Infrastructure Management Services (IMS), and Deployment Management Services (DMS).
- Hardware resources that consist of computer system, underlay network, and hardware accelerator manager.

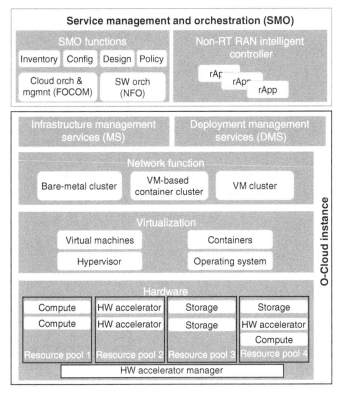

Figure 7.2 Cloud architecture.

- Operating system (OS) that manages and abstracts the computer system's hardware and software resources and provides common services for network functions.
- Virtualization layer that houses hypervisor (offering virtual machines, VM, as separated computer systems) and container engine that is an OS to offer name spaces, quotas, and management for containers.
- NF layer that includes bare meal container cluster (network-connected computer systems supporting containers in cluster configuration), VM-based container cluster (network-connected VMs), and VM cluster (network-connected computer systems).
- O-Cloud images repository.

In the cloud deployment scenarios, the main stakeholders are operator, cloud service provider (CSP), vendor, and third party such as service provider (SP) and vertical representative.

7.3.2.6 O-RAN Study on Security for Application Lifecycle Management

Lifecycle management of applications that are deployed in O-RAN follows standard SDLC models of which Figure 7.3 presents an example. The application lifecycle phases are development, onboarding, and operation as per the O-RAN Alliance's O-RAN OAM Architecture description.

The study on security for application lifecycle management presents key issues and potential solutions based on the reported findings.

The issues include the development of application package, onboarding and deployment of application, and operation and maintenance of application. The suggested solutions relate to application package authenticity and integrity protection, secure update of applications, descriptor of application security, secure logging and monitoring of applications, and SW bill of material (BoM) requirements for applications.

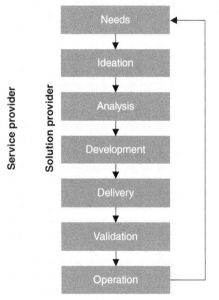

Figure 7.3 Software development lifecycle process.

7.3.2.7 O-RAN Study on Security Log Management

The O-RAN study on security log management describes the concept and assumptions of secure log management and outlines some of the most relevant, related issues, such as log authentication attempts, logging access to personal data, emergency calls related logs, logging details of O-Cloud, xApps, rApps, and VNF/CNF. These key issues can be categorized to security log contents (payload), security log transfer mechanisms, and security logs formats and schemas.

Figure 7.4 depicts the components of O-RAN architecture from the security log management perspective.

Log-management consists of application-level normal-logs, the authentication-level audit-logs, and the OS-level logs, that is, system logs. The log collector provides the log data to a log-data repository which can be accesses via graphical user interface (GUI) in order to visualize the data. The log data is also available for the security information and event management (SIEM) data aggregation tool. Intrusion detection system (IDS) and intrusion prevention system (IPS) can use the aggregated data.

7.3.2.8 O-RAN Study on Security for Service Management and Orchestration

SMO can have security risks in its internal functions as well as in internal and external interfaces. As it manages services and orchestration and has deep visibility into the infrastructure, it is thus of utmost importance to secure it. Otherwise, an SMO security vulnerability could lead to exposure of an entry point for attacks against O-RAN components and lateral movement across O-RAN.

The O-RAN study on security for service management and orchestration (SMO) introduces the SMO architecture, functions, applications, interfaces, and information. The documents subsequently present threat model and template, potential threats and exploits, and specifically SMO threats. The document also outlines threat analysis, security controls, and

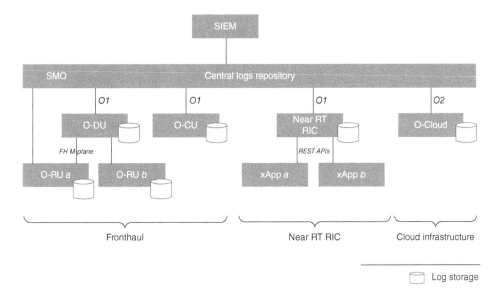

Figure 7.4 Security log management in O-RAN.

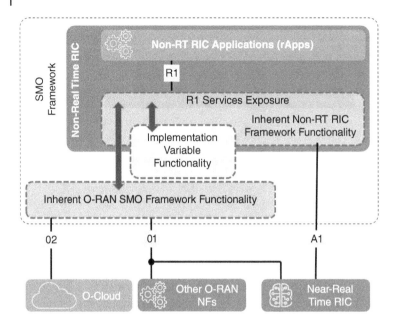

Figure 7.5 SMO architecture. Source: Reproduced from O-RAN Alliance.

risk assessment in general SMO environment, in O2 interface, and in external interfaces. The document finally discusses primary security issues and presents recommendations to cope with the threats.

The protected assets include the actual SMO framework and functions, as well as Non-RT RIC, the internal communications of SMO, O1, O2, R1, and A1 interfaces, Open Fronthaul M-plane, external interfaces, user management interfaces, PKI, logging, data, and AI/ML models, algorithms, and methods.

Figure 7.5 depicts the O-RAN SMO architecture. As identified in this figure, the risk analysis should consider A1, R1, O1, and O2 interfaces. In addition these, OFH M-plane and external interfaces should be included, although not shown in the figure. For the SMO data and information, the risk analysis should consider O1, O2, and AI/ML data and secret stores (X.509 certificates, private keys, and administrator credentials).

7.3.2.9 O-RAN Study on Security for Shared O-RU

The O-RAN study on security for shared O-RU (Shared ORU) presents shared O-RU architecture, functions, interfaces, and information and outlines respective threat model and template, potential threats and exploits, and shared O-RU threats. The document continues presenting threat analysis, security controls, risk assessment, primary security issues, and gives recommendations.

The presented steps of the threat modeling process are

- Identify assets of the shared O-RU that need protection.
- Identify threats that potentially impact the shared O-RU and that may use it to impact other O-RAN components.

- Identify the attack surface and attack vectors in the shared O-RU.
- Measure risk related to confidentiality, integrity, and availability.
- Recommend management, operational, and technical security controls to protect the confidentiality, integrity, availability, and information of the shared O-RU.

As for the functions, the shared O-RU risk analysis should consider the actual, shared O-RU, O-DU Host and Tenant, SMO Host and Tenant, and Neutral Host Controller. The interfaces to be considered are O1, OFH M-Plane (OAM Config and Carrier Config), and SMO host–tenant interface.

7.3.2.10 O-RAN Study on Security for Near-Real-Time RIC and xApps

The study on security for near real-time RIC and xApps (micro service-based applications) outlines the RIG concepts and summarizes related key issues, mitigations, and solutions. The document considers the security aspects of the near-RT RIC platform and xApps, as well as the associated network and management interfaces (E2, A1, O1) and APIs.

The issues include malicious xApps deployed on the near-RT RIC, compromised and conflicting xApps, isolation between xApps, compromise of ML data pipelines and altering the ML learning models in near-RT RIC. The document also identifies potential threats on data protection and privacy in near-RT RIC database, onboarding untrusted xApps, and xApp exploitation via malicious A1 policies, as well as vulnerabilities in near-RT RIC API authentication and authorization, and xAppID abuse.

Figure 7.6 present the near-RT RIC architecture.

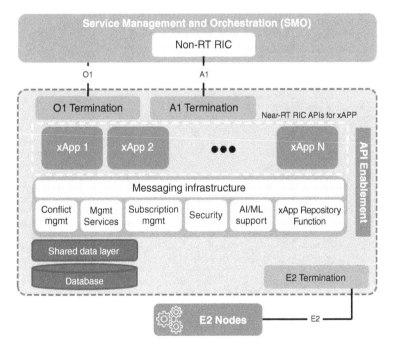

Figure 7.6 Architecture of Near-Real-Time RIC. Source: Reproduced from O-RAN Alliance.

As indicated in Figure 7.6, the near-RT RIC interfaces are

- E2 between the near-RT RIC and the E2 nodes (O-DU, O-CU-UP, O-CU-CP, O-eNB);
- O1 between the SMO and O-RAN managed functions (near-RT RIC, O-RU, O-DU, O-CU-UP, O-CU-CP, O-eNB); and
- A1 between the non-RT RIC at the SMO and the near-RT RIC.

Figure 7.7 summarizes the APIs of the near-RT RIC. These APIs have the following tasks:

- *Non-RT RIC Related APIs (also A1 APIs)* allow to access A1-related functionality.
- *O-DU/O-CU Related APIs (also E2 APIs)* allow to access E2-related functionality and associated xApp subscription management and conflict mitigation functionality.
- SMO-Related APIs (also Management APIs) allow to access management-related functionality for services such as O1 termination, management services and logging, tracing, and metrics collection.
- ML model-related APIs.
- FACPS-related APIs.
- SDL API allows to access shared data layer-related functionalities over R-NIB, UE-NIB, and other data in the database.
- Enablement APIs are between xApps and API enablement functionality.

7.3.2.11 Complementing Material

O-RAN Threats and Remediation Analysis V2.0 has identified assets that need to be protected. It has also analyzed O-RAN component vulnerabilities and respective, potential

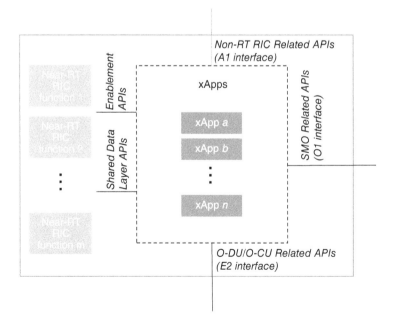

Figure 7.7 Near-RT RIC-related APIs. The near-RT RIC houses the framework for transactions on APIs.

threats. It outlines security principles to guide stakeholders in their efforts to set up a secure end-to-end O-RAN system.

7.3.3 Industry and Government Entities

7.3.3.1 Open RAN MoU

In 2021, Vodafone Group, Deutsche Telekom AG, Orange S.A., Telefónica S.A., and TIM S.P.A. founded the Open RAN Memorandum of Understanding (MoU) to support the roll-out of Open RAN for future mobile networks [13]. The MoU representatives of the O-RAN Alliance support development of Open RAN specifications so that they fortify compliance with security and data privacy requirements in the EU to allow mobile network operators (MNOs) to build solutions that meet security and legislative requirements.

7.3.3.2 NIS

EU published in 2022, together with NIS Cooperation Group, an assessment of Open RAN security. The aim of this effort is to advance the development of adequate security controls for specification, development, procurement, system integration, testing, and operations.

The NIS Cooperation Group has stated that Open RAN will have a significant impact on a number of risks that were already identified in the EU coordinated risk assessment of 5G networks published in October 2019. Furthermore, NIS has identified several new risks and vulnerabilities introduced by Open RAN stating that if not adequately mitigated, these risks could have a significant, negative impact on the overall security of 5G deployments using Open RAN [15].

7.3.3.3 EU

In the EU coordinated risk assessment of 5G networks from 2019, the NIS Cooperation Group has identified a number of categories of risks with nine concrete risks that are of strategic importance from the EU perspective. These risks remain relevant, to various extents, to Open RAN deployments, and can evolve in Open RAN networks [15].

7.3.3.4 FCC

FCC has released a report on challenges to the development of Open RAN technology and recommendations on how to overcome them [16]. The key weight of this report is on identifying issues facing the deployment of Open RAN equipment in the United States, but it also lists key threats in global environment.

The authoring entity of the report, CSRIC VIII, has summarized recommendations to advance security, reliability, and interoperability of Open RAN equipment in the United States. It also has listed suggestions on efforts that could be undertaken to support secure and interoperable Open RAN design and deployment in the new environment that enables mobile network operators to rely on equipment from multiple vendors and still needing to ensure interoperability and security.

In this report, the key items under analysis are on how to secure management and orchestration, RAN intelligent controller (RIC), open Fronthaul interface, O-Cloud, and O-RAN coexistence orchestration to mitigate security vulnerabilities in multi-RAT deployment environment. It also summarizes recommendations for Open RAN architecture.

The Open RAN architecture includes important changes through disaggregation, which opens the ecosystem for increased vendor diversity. The changes are related to the O-RAN lower layer split (LLS) 7-2x OFH interface, RIC, and RAN applications (rApp and xApp). Despite of the benefits in the diversification of the vendor environment, the new interfaces and network functions of the O-RAN potentially expose new attack surfaces. As concluded in [16], security controls at each layer are needed to protect the network functions, interfaces, and data from external and internal threats. The WG11 of the O-RAN Alliance has evaluated threats in this new environment to develop further specifications to help operators mitigate the risks. The base assumption of the O-RAN Alliance is the zero trust architecture (ZTA) as per US NIST SP 800-207 to serve as protection from external and internal threats.

Management and orchestration as per O-RAN are done via the new SMO function for taking care of the 3GPP Network Manager functions by introducing standardized interfaces. Securing the SMO and its associated interfaces is of utmost importance because compromising it, the bad actors can obtain access to also other Open RAN components and provide means for lateral movement across Open RAN putting in risk the RAN confidentiality, integrity, availability, and authenticity, as well as internal and external data stores the SMO accesses through APIs.

RIC can rely on rApps and xApps to enhance RAN optimization leveraging AI and ML. This means that the security related to AI and ML data and models is important. In addition, rApp can interface with other rApps through R1 interface to enhance automated decisions. To protect this part, secure peering between rApps is important involving also mutual authentication across the R1 interface. To protect the confidentiality and integrity, R1 interface must be secured against potential malicious rApps that may intend to snoop and inject messages on this interface.

Open Fronthaul Interface's M-Plane is end-to-end secured, and it enables O-RU controllers to manage the O-RU operations. For the respective O-RU controller's configuration and data recovery, the NETCONF protocol and YANG models are in use. For the protection of NETCONF, it must be securely transported over SSH v2 or TLS v1.2, or newer versions, and the authentication on M-Plane must be compliant with the O-RAN requirements such as username/password, public key certificates, or Public Key Infrastructure involving X.509 certificates in the case of SSH v2, or by using PKI X.509 certificates in the case of the use of TLS. The OFH can be deployed in a hierarchical or hybrid mode, and both need to be shielded against bad actors. In the hierarchical model, it is possible to isolate the fronthaul traffic within the interface between the O-RU and O-DU interface by applying segmentation. In the hybrid model, the traffic of the M-Plane may connect other centralized management systems, which could require additional security measures [17].

O-Cloud in the Open RAN architecture is visible for the SMO via the O2 interface and is an entity located logically on top to serve Open RAN functions and to run applications. The O-Cloud software and interface needs to be thus sufficiently well protected to shield against any attacks against the cloud infrastructure aiming to compromise confidentiality, integrity, availability, and authenticity of the Open RAN environment. One of the ways to secure the O-Cloud is to rely on secure hardware that uses hardware security module (HSM) to establish a hardware root of trust.

O-RAN coexistence orchestration in multi-RAT environment may be challenging because multi-RAT may increase exposure of the network's attack surface to malicious actors. As Open RAN decouples the software from the hardware and moves orchestration to a cloud-native environment, orchestration can become automated and thus untrusted. Also, the vendor-agnostic RATs may create unknown attack surface between multi-vendor solutions.

7.4 New Threats

7.4.1 Overview

This section discusses treats and attack vectors on cloud / virtualized environment and presents guidelines for protection.

7.4.2 Machine Learning

AI and ML can be of use in 5G, both to manage the highly complex functions and to optimize the network's performance, as well as to offer fluent user experiences at the application level. As reference [18] points out, ML systems offer high flexibility in dealing with evolving input in a variety of applications, but at the same time, ML algorithms can also be a target of attack by a malicious adversary.

In specifically 5G, the ML can serve as an additional component to optimize functions such as network slicing orchestration. It is an example of the new functions that are highly dynamic and can tackle environments that change much faster than has been typical in legacy systems. The optimal performance thus benefits from the self-learning methods that improve the outcome much more efficiently than would be possible adjusting parameters manually.

The ML models and algorithms require, though, adequate, initial training to work well. This training period may turn out to be critical also for the security of the systems it is integrated to. An example of badly trained ML system can be a chatbot that learns inappropriate language and starts using it increasingly with the customers it is serving.

Reference [19] points out that research shows many ML systems are vulnerable to adversarial attacks, imperceptible manipulations that cause models to behave erratically. This reference presents an example of vulnerabilities in ML models served over the web such as API services of large tech companies, quoting Alex Polyakov, co-founder and CEO of Adversa, on most of the models online being vulnerable proven by several research reports, making it possible to train an attack on one model and then transfer it to another model without knowing any special details of it. Another related scenario, according to this source, is to perform CopyCat attack to steal a model, apply the attack on it, and then use this attack on the API.

APIs are becoming more common in ICT environment including also 5G so this type of potential attack vector through ML can be relevant to many cases that are not yet deployed.

7.4.3 Open Interfaces

Virtualization refers to an abstraction layer on top of the hardware, which is a hypervisor on which software runs. Cloudification, in turn, facilitates the software to run in a cloud infrastructure instead of needing to run in locally [20].

Along with this evolution, the more efficient diversification of the radio access network equipment vendors has become reality. As stated in [15], in the traditional model, MNOs need to obtain the complete RAN from a single supplier or rely on different (competing) vendors that provide RAN merely in their dedicated and separated geographic areas. This strategy has been implemented because in practice, there are issues in the interoperability of the RAN vendors along with different phases each one support as for the standard features. Also, each vendor tends to have proprietary solutions.

The key benefit of Open RAN is to optimize this environment allowing RAN components of several suppliers via interoperable interfaces. In this way, when the equipment of diverse sources is deployed in the same geographic area, they work regardless of the vendor thanks to open and standardized interfaces. The work of Open RAN is in fact based on this specific strategy, development of open interfaces. Other important enablers of Open RAN include the splitting or disaggregating of network functions, and the cloudification and virtualization of network functions [15].

Regardless of the clear benefits of the open interfaces, they also may have unknown aspects that, in the worst case, can open completely new threat vectors. On the EU wide initiatives, reference [21] presents risk assessment of 5G networks security informing a publication about a high-level report on the coordinated risk assessment of 5G networks. Of the key statements of this document, it emphasizes the fact that the move to software and virtualization through software defined networking (SDN) and network functions virtualization (NFV) will represent a major shift from traditional network architecture as functions will no longer be built on specialized hardware and software as the functionality and differentiation will take place in the software. The reference states that from a security perspective, "this may bring certain benefits by allowing for facilitated updating and patching of vulnerabilities" and that "at the same time, such increased reliance on software, and the frequent updates they require, will significantly increase the exposure to the role of third-party suppliers and the importance of robust patch management procedures." This report and the work are tightly related to the EU initiative on cybersecurity of 5G networks that introduces EU Toolbox of risk-mitigating measures [22].

7.4.4 Open-Source SW

As stated by GSMA, 5G network infrastructure is of utmost importance for society and the economy via its support of wide range of mobile connectivity services. Along with the more open interfaces and their facilitation to mix and match new stakeholders, the 5G landscape will be increasingly complex and has raised questions about potential security vulnerabilities in, e.g. the 5G supply chain. This is due to the new way of RAN disaggregating the network components which, at the same time, can potentially broaden the number of threat surface for any network. As GSMA states, this trend is not only to raise security concerns but it also provides mobile operators with more possibilities to procure and ensure the up-to-date configuration of adequately secure, reliable, and resilient networks in align with best practice security principles. These security measures would be additional to the overall evolutions of the hardened security architecture specifications of 5G [23].

5G relies thus increasingly on open-source software (OSS). As the NIST states, the OSS can be accessed, used, modified, and shared by anyone and is often distributed under licenses that align with the definitions of the Open Source Initiative and/or the Free Software Foundation.

OSS may be exploited regardless of its advantages and strengths. The Synopsis 2020 Open Source Security and Risk Analysis Report had reasoned that 99% of codebases had open source components of which 49% had high-risk vulnerabilities, meaning that the OSS could expose attack vectors such as intentional backdoors by malicious developers, propagation of vulnerabilities through reuse, exploitation of publicly disclosed vulnerabilities, and human error [24]. Reference [24] further reminds that the Heartbleed vulnerability with OpenSSL was one of the major examples of the OSS vulnerabilities in cryptographic software tasked to handle big amount of we traffic. In this critical environment, Heartbleed could expose private keys and other sensitive information.

To overcome the challenges with the OSS, Reference [24] suggests that its protection requires an SDLC process as described by the US-CERT. The hardened security of the software benefits from secure coding practices, testing, and code audits as well as the overall developer education. Also, the SDLC benefits from the inclusion of supply chain management, asset inventory, license compliance, and patch management [25].

Reference [24] also reminds that it is challenging to review and audit all the essential in the highly complex, modern open-source ICT ecosystem. Thus, even if code audits are needed, the visibility to the source code of products would not automatically guarantee its security. It is important to use up-to-date code testing methods for revealing security vulnerabilities.

Relevant organizations can assist in finding ways to implement the security controls in place, such as the US Department of Commerce's National Institute of Standards (NIST), The Linux Foundation (LF), and its LF Core Infrastructure Initiative (CII), as well as OWASP that provides automated vulnerability detection tools. Of the concrete guidelines, the NIST publication on "Secure Software Development Framework (SSDF) Version 1.1" presents recommendations for mitigating the risk of software vulnerabilities [26].

7.4.5 Supply Chain

The EU Toolbox has highlighted that significant risks can originate from the supply chain of 5G networks stating that more suppliers available in the network could lead to exposure to a higher number of supply chain risks [15]. The reference reminds that the risk profile of each supplier can be assessed based on the criteria recommended in the EU coordinated risk assessment and is an important source of identified vulnerabilities.

One of the key publications related to supply chain risk mitigation is the NIST SP 800-161, revision 1, on Supply Chain Risk Management Practices for Federal Information Systems and Organizations" that provides guidance for OSS evaluation and remediation [27]. It serves as a guidance to organizations on identifying, assessing, and mitigating cybersecurity risks throughout the supply chain. It considers cybersecurity supply chain risk management (C-SCRM) and their breakdown into risk management activities by applying a multilevel, C-SCRM-specific approach, including guidance on the development of C-SCRM strategy implementation plans, C-SCRM policies, C-SCRM plans, and risk assessments for products and services.

7.4.6 Misconfiguration

Misconfiguration of networks can open unpredictable, new attack vectors to Open RAN environment. This can be especially challenging to cope with along with the integration of increasing number of components provided by different suppliers. Also, the virtualization management, being new way of work, if not fully understood by the operating entities, can be a root of issues. Other issues include the still less mature MNO processes for the Open RAN deployment lifecycle that may set challenges to the network configuration, increasing the risk of misconfigured networks that result in faults and potential security holes.

Also, the developing standards may impact the network design lessening the optimal architectures. The software presents a complexity challenge through the variety of their components, libraries, and other aspects increasing the permutation of the combinations with the respective hardware on which they run. Especially the integration of the required security features may not be straightforward and might result in security vulnerabilities when these are deployed less optimally, or if there is a lack of important protection features [15].

7.4.7 Low Product Quality

Product quality may be negatively impacted in Open RAN environment along with the increased number of components from different suppliers. This, in turn, can augment the security risks – as is the case in the weakest link of the chain. As the Open RAN concept is still in its initial phase, and the research and development of ideas needs to be prioritized for achieving fast time-to-market of the products, it is to be seen how the quality may vary in practical products and deployments in scenarios involving interconnected and interoperating components from increased number of vendors [15].

7.4.8 Lack of Access Controls

The Open RAN infrastructure needs to be accessed for procedures of maintenance and operation, and as more players will be involved in the multi-vendor environment, each requiring their maintenance access (e.g. for configuration and fault management, as well as performance monitoring), this can also open up security vulnerabilities if not managed adequately. As the initial phase of the Open RAN deployment may be very different and complex from the more traditional architectures from which MNOs have already more experiences, the new players may involve also new third-party representatives and system integrators as the equipment vendors and MNOs may outsource certain functions [15].

7.4.9 Other Risks

In addition to the risks outlined in the previous sections, reference [15] identifies a variety of other potential risks such as:

- Exploitation of 5G networks by organized crime groups (OCGs) to enter and disrupt the network.
- Disruption of critical infrastructures or services via the increased number of suppliers and thus new open interfaces, for the actors taking advantage of the still developing stage of Open RAN and NFV security controls. This could put the critical national infrastructures relying on 5G networks as a target of the bad actors.
- Interruption of electricity supply and other support systems may jeopardize the functioning of telecom infrastructure, including Open RAN architecture. To prevent major impacts, adequate power supply and battery backup must be in place in the most critical sites.
- IoT exploitation via connected terminals may be used to attack Open RAN components and their network functions as well as virtualized multi-access edge computing applications.
- Open RAN functions and interfaces threat surface expands due to an increased number of suppliers, components, and interfaces, and example of this is the fronthaul interface exploitation to carry out denial-of-service attacks, interception or tampering attacks to compromise availability, confidentiality, or integrity.
- Open RAN will provide access to new third parties in terms of previously hidden information flows, so that this data, even if the actual contents, may be protected, it could possibly reveal related information such as real-time location of users.
- The multi-vendor setup, even it is desired, can also increase the interconnection complexity when there is a need to narrow down the root cause of issues and resolve network malfunction. This may result in increased time for recovery along with multiple suppliers needing to provide simultaneously remote support.
- Still during the evolving phase, O-RAN specifications may lack the final security measures leading to insecure RAN products.
- Open RAN deployments rely on the use of cloud, which may lead to a risk of increased dependency on a small number of cloud service and infrastructure providers, which in turn could lead to supplier lock-in. The centralization of such providers could mean that, upon possible vulnerability materializing, the damage can be significant.
- The integration of several components and their versions from multiple suppliers may not turn about to be seamless and may expose security vulnerabilities and result in performance level degradation.
- Resource sharing must be well protected in Open RAN along with virtualized network functions running on the same hardware, to prevent potential other network functions security issues impacts.
- Misconfiguration of the virtualized network functions and software may lead to new and hard-to-predict vulnerabilities, for example, in the hypervisor, or in SMO functions.

More information of the variety of vulnerability aspects can be found in Refs. [28–32].

7.5 O-RAN Interface Protection Aspects

7.5.1 General

There are various case studies of the O-RAN-specific weaknesses and protection mechanisms. Reference [15] summarizes some of the key risks that are amplified or brought by Open RAN including:

- There will be an increased number of entry points for malicious actors regardless of the supplier along with the larger number of suppliers and their components, which in turn may expose expanded threat surface and create a more complex environment increasing vulnerability risks or risks for failure. This may result in unwanted data and information flow to new third-party applications.
- Increased risk to misconfigure networks because the respective new specifications may not be sufficiently mature and secure by design.
- Along with the outsourcing of functions and offloading the data to external cloud environment, there will be increased dependency on cloud service and infrastructure providers.
- Network resource sharing can also increase risks and impact on other network functions if sufficient controls are not considered.

The security of Open RAN environment needs to be thus based on secure communications between network functions within the Open RAN deployment ecosystem. The communications endpoints need to be authenticated based on trust including trusted certificate authorities (CA) for the provisioning of identity [33].

7.5.2 Protection of Interfaces

As stated in [33], the architecture by O-RAN Alliance is by nature open and secure for its interfaces between all the involved components. The cryptographical procedures of all the communications on the Open RAN interfaces cover encryption, integrity, and replay protection. Figure 7.8 depicts the Open RAN security architecture in 5G domain. The SMO refers to the SMO framework.

The security aspects of the involved interfaces are summarized later (Tables 7.2 and 7.3).

7.5.3 Mutual Authentication

As stated in [33], mutual authentication is applied between two entities for setting up a secure encrypted connection between them. The reason for mutual authentication is to provide means to shield against potential rogue NFs or xAPPs that may otherwise jeopardize the network security.

Mutual authentication can be based on operators' X.509 certificates, while secure connections are established based on IPsec and TLS protocols. In Open RAN, all network elements including O-CU-CP, O-CU-UP O-DU, and O-RU support X.509 certificate-based authentication and related features such as automatic enrollment and re-enrollment with an operator CA server using a protocol such as enrollment over secure transport (EST) or 3GPP-based CMPv2 [34, 35]. Also, the xAPPs in the near-RT RIC are securely on-boarded

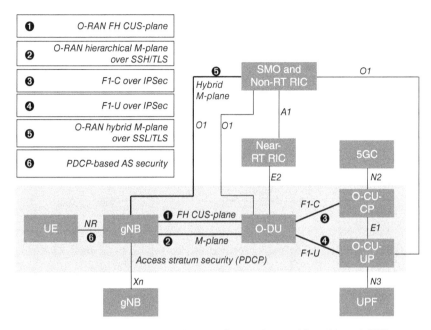

Figure 7.8 Open RAN security architecture. Source: Adapted from Mavenir [33].

Table 7.2 Security aspects of Open RAN as per 3GPP.

Interface	Description	Security
E1	O-CU-CP – O-CU-UP	IPSec/DTLS
Xn	Source gNB – target gNB	IPSec/DTLS
Backhaul	N2 and N3	IPSec/DTLS
Midhaul	F1-C and F1-U	IPSec/DTLS

Source: Adapted from Mavenir [33].

similarly to other microservices. According to [33], the O-RAN Alliance is expected to use CA signed X.509 certificates to authenticate before communicating over E2.

7.5.4 Security Aspects for Near-Real-Time RIC

As stated in [33], the near-real-time RIC (near-RT RIC) is software defined networking component that houses xApps, i.e. third-party extensible microservices. The task of the xApps is to take care of radio resource management (RRM) services for those NFs that gNB has managed internally prior to Open RAN.

The near-RT RIC has an open E2 interface, standardized by the O-RAN, for the communications with the O-CU-CP, O-CU-UP, and the O-DU. In addition, the near-RT RIC also connects, through A1 and O1 interfaces, to non-RT RIC and the SMO framework.

Table 7.3 Security aspects of Open RAN as per O-RAN (all defined by WG1 STG except the Open FH M-plane, which is defined by WG4).

Interface	Description	Security
Open FH, M-plane	O-RU – O-DU/SMO	SSHv2, TLS
Open Fronthaul (CUS-plane)	O-DU – O-RU	TBD
O1	SMO – O-RAN management	TBD
E2	Near-RT RIC (xApps) – O-CU-CP	TBD
A1	Near-RT RIC – non-RT RIC	TBD
O2	SMO – O-Cloud	TBD

Source: Adapted from Mavenir [33].

Reference [33] has identified the following security aspects related to the near-RT RIC:

- Secure E2 interface between the near-RT RIC and the O-CU-CP, O-CU-UP, and O-DU.
- Conflict resolution and xApp authentication.
- User identification within the near-RT RIC.

7.5.5 Security Aspects of Non-Real-Time RIC

As summarized in [33], the O-RAN non-RT RIC is designed for the non-real-time control of the RAN that takes place through declarative policies and objective intents. The non-RT RIC has the following characteristics:

- It is deployed in an SMO framework.
- It provides declarative policy guidance for cell-level optimization via optimal cell parameter configuration values though O1.
- It provides declarative policies for UE-level optimization to the near-RT RIC through A1.
- It translates the recommended declarative policy from the non-RT RIC over A1 into per-UE control and imperative policy over E2.
- It develops ML/AI-driven models for policy guidance and non-RT optimization as rApp microservices. The rApps communicate with the xApps through A1 in order to optimize procedures and functions of the underlying RAN.

As summarized in [33], the Non-RT RIC provides a secure interface between non-RT RIC and the O-CU-CP, O-CU-UP, and O-DU. It also provides conflict resolution between the non-RT RIC and the O-CU-CP, O-CU-UP, and O-DU.

7.5.6 Trusted Certificate Authorities

As stated in [33], the CA should be ideally audited under the AICPA/CICA WebTrust Program for Certification Authorities. This would promote confidence and trust in the CA servers in Open RAN for authenticating network elements.

References

1 O-RAN Alliance, "O-RAN Use Cases and Deployment Scenarios," O-RAN Alliance, February 2020.

2 O-RAN Alliance, "O-RAN: Towards an Open and Smart RAN," O-RAN Alliance, October 2018.

3 Rakuten, "The definite guide to Open RAN security," Rakuten, October 2022. [Online]. Available: https://assets.website-files.com/6317e170a9eabbe0fbbf4519/63582c8cec69a24b2bcde588_221025-Security-Handbook.pdf. [Accessed 14 December 2022].

4 ETSI, "TSI White Paper No. 23: Cloud RAN and MEC: A Perfect Pairing," ETSI, February 2018. [Online]. Available: https://www.etsi.org/images/files/ETSIWhitePapers/etsi_wp23_MEC_and_CRAN_ed1_FINAL.pdf. [Accessed 30 August 2020].

5 M. Shelton, "Open RAN Terminology," Parallel Wireless, 20 April 2020. [Online]. Available: https://www.parallelwireless.com/open-ran-terminology-understanding-the-difference-between-open-ran-openran-oran-and-more/. [Accessed 29 July 2020].

6 Open RAN Policy Coalition, "Open RAN security in 5G," Open RAN Policy Coalition, April 2021. [Online]. Available: https://www.openranpolicy.org/wp-content/uploads/2021/04/Open-RAN-Security-in-5G-4.29.21.pdf. [Accessed 14 December 2022].

7 Telefonica, "Major European operators accelerate progress on Open RAN," Telefonica, 21 February 2023. [Online]. Available: https://www.telefonica.com/en/communication-room/major-european-operators-accelerate-progress-on-open-ran/. [Accessed 28 Mach 2023].

8 Telecom Infra Project, "OpenRAN MoU Group Page," TIP, [Online]. Available: https://telecominfraproject.com/openran/. [Accessed 1 June 2023].

9 Telecom Infra Project, "OpenRAN 5G NR Base Station Platform Requirements Document v5.0," TIP, 2020.

10 Techplayon, "Open RAN (O-RAN) Reference Architecture," Techplayon, 28 October 2018. [Online]. Available: http://www.techplayon.com/open-ran-o-ran-reference-architecture/. [Accessed 10 August 2020].

11 "Telefónica views on the design, architecture, and technology of 4G/5G Open RAN networks," Telecom Infra Project, January 2021.

12 O-RAN Working Group 11, "Security Requirements Specifications," O-RAN Alliance, March 2023.

13 Deutsche Telekom, Orange, Telefonica, TIM, Vodafone, "Open RAN Security White Paper: Under the Open RAN Mou," 2021. [Online]. Available: https://cdn.brandfolder.io/D8DI15S7/at/45zqtkzjp4n9ncn77mkqbx8/Open_RAN_MoU_Security_White_Paper_-_FV.pdf. [Accessed 14 December 2022].

14 O-RAN Alliance, "O-RAN specifications," O-RAN Alliance, 2022. [Online]. Available: https://www.o-ran.org/specifications. [Accessed 14 December 2022].

15 NIS Cooperation Group, "Report on the cybersecurity of Open RAN," NIS Cooperation Group, 11 May 2022. [Online]. Available: https://d110erj175o600.cloudfront.net/wp-content/uploads/2022/05/11160610/OPEN.pdf. [Accessed 14 December 2022].

16 FCC, "Report on challenges to the development of ORAN technology and recommendations on how to overcome them," Communications security, reliabilty, and inyeroperability council VIII, 2022.

17 K. Murphy, "Centralized RAN and Fronthaul," Ericsson, [Online]. Available: https://www.isemag.com/wp-content/uploads/2016/01/C-RAN_and_Fronthaul_White_Paper.pdf. [Accessed 6 July 2019].

18 B. N. R. S. A. D. J. J. D. T. Marco Barreno, "Can Machine Learning Be Secure?," Computer Science Division, University of California, Berkeley, 21-24 March 2006. [Online]. Available: https://people.eecs.berkeley.edu/~adj/publications/paper-files/asiaccs06.pdf. [Accessed 18 February 2023].

19 B. Dickson, "Machine learning security vulnerabilities are a growing threat to the web, report highlights," The Daily Swig, 30 June 2021. [Online]. Available: https://portswigger.net/daily-swig/machine-learning-security-vulnerabilities-are-a-growing-threat-to-the-web-report-highlights. [Accessed 18 February 2023].

20 RedHat, "What's the difference between cloud and virtualization?," RedHat, 7 March 2018. [Online]. Available: https://www.redhat.com/en/topics/cloud-computing/cloud-vs-virtualization. [Accessed 15 March 2023].

21 EU, "EU-wide coordinated risk assessment of 5G networks security," European Union, 19 October 2019. [Online]. Available: https://digital-strategy.ec.europa.eu/en/news/eu-wide-coordinated-risk-assessment-5g-networks-security. [Accessed 15 March 2023].

22 EU, "Cybersecurity of 5G networks - EU Toolbox of risk mitigating measures," European Union, 29 January 2020. [Online]. Available: https://digital-strategy.ec.europa.eu/en/library/cybersecurity-5g-networks-eu-toolbox-risk-mitigating-measures. [Accessed 15 March 2023].

23 GSMA, "Open and Virtualised Radio Access Networks: An Explanatory Guide for Policymakers," GSMA, February 2021. [Online]. Available: https://www.gsma.com/publicpolicy/wp-content/uploads/2021/02/GSMA_Open_and_Virtualised_Radio_Access_Networks_An_Explanatory_Guide_for_Policymakers.pdf. [Accessed 18 February 2023].

24 R. B. T. T. Scott Poretsky, "Open source software security in an ICT context – benefits, risks, and safeguards," Ericsson, 14 January 2021. [Online]. Available: https://www.ericsson.com/en/blog/2021/1/open-source-security-software. [Accessed 18 February 2023].

25 KeySight, "O-RAN Next-Generation Fronthaul Conformance Testing (whitepaper)," KeySight, USA, 22 July 2020.

26 K. S. (. C. D. D. Murugiah Souppaya (NIST), "Secure Software Development Framework (SSDF) Version 1.1: Recommendations for Mitigating the Risk of Software Vulnerabilities," NIST, February 2022. [Online]. Available: https://csrc.nist.gov/publications/detail/sp/800-218/final. [Accessed 18 Februsry 2023].

27 A. S. N. B. K. W. A. H. M. F. Jon Boyens, "Cybersecurity Supply Chain Risk," NIST, May 2022. [Online]. Available: https://nvlpubs.nist.gov/nistpubs/SpecialPublications/NIST.SP.800-161r1.pdf. [Accessed 18 February 2023].

28 Federal Office for Information Security (BSI), "Open RAN Risk Analysis," Federal Office for Information Security (BSI), 21 February 2022. [Online]. Available: https://www.bsi.bund.de/SharedDocs/Downloads/EN/BSI/Publications/Studies/5G/5GRAN-Risk-Analysis.pdf?__blob=publicationFile&v=7. [Accessed 15 March 2023].

29 ENISA, "NFV Security in 5G - Challenges and Best Practices," ENISA, 24 February 2022. [Online]. Available: https://www.enisa.europa.eu/publications/nfv-security-in-5g-challenges-and-best-practices. [Accessed 15 March 2023].

30 EU, "EU strategic dependencies and capacities: second stage of in-depth reviews," EU, 22 February 2022. [Online]. Available: https://ec.europa.eu/docsroom/documents/48878/attachments/2/translations/en/renditions/native. [Accessed 15 March 2023].

31 Fujittsu, "A brief look at O-RAN security," Fujitsu, 2022. [Online]. Available: https://www.fujitsu.com/global/documents/products/network/Whitepaper-A-Brief-Look-at-O-RAN-Security.pdf. [Accessed 14 December 2022].

32 United States Government, "Open Radio Access Network Security Considerations," United States Government, [Online]. Available: https://arxiv.org/abs/2101.10856. [Accessed 14 December 2022].

33 Mavenir, "Security in Open RAN," Mavenir, January 2021. [Online]. Available: https://gsma.force.com/mwcoem/servlet/servlet.FileDownload?file=00P6900002qUScBEAW. [Accessed 14 December 2022].

34 ITU-T, "GSTR-TN5G: Transport network support of IMT-2020/5G," ITU, February 2018. [Online]. Available: https://www.itu.int/dms_pub/itu-t/opb/tut/T-TUT-HOME-2018-PDF-E.pdf. [Accessed 30 August 2020].

35 M. Bougioukos, "Preparing microwave transport network for the 5G world," Nokia, 2017.

8

Open RAN Deployment Considerations

8.1 The Evolution of the RAN Deployment Strategy

In early generations of the mobile communication systems, the goal of the mobile network operator was primarily to provide a large, contiguous service area (coverage) for basic voice services. This goal was achieved by installing complete base station systems across the territory which the operator desired to cover and connecting each base station directly to the core network. Over time, with the evolution of the services that an operator provided as well as the growth of the users, the basic voice coverage became only adequate for rural and sparsely populated areas. As data services became more popular, providing coverage was just a necessity and the focus shifted instead to how the RAN could offer the capability of supporting higher traffic volumes and higher throughput. Having more capacity to serve more users became a key design principle of the RAN especially in urban and suburban areas.

While the additional spectrum that became available to mobile operators over time and improvements in the radio technology enabled more efficient use of the spectrum, it became apparent that the deployment strategy needed to evolve from simply adding as many full base station systems as possible to more efficient and cost-effective deployments.

The introduction of centralized RAN discussed in Chapter 3 enabling the decoupling of the elements of the base station provided the first opportunity to experiment with optimized deployment whereby, rather than utilizing a full base station, the same baseband unit (BBU), residing in a data center, could be used to serve multiple cell sites only consisting of the remote radio head (RRH) and antenna system. This solution, which evolved into the so-called virtualized radio access network (vRAN) architecture, while suitable for meeting certain deployment criteria suffered from limited flexibility in distributing functions between the cell site and the data center. Furthermore, centralized RAN and in many instances vRAN requires the operator to procure the components of the base station, that is, BBU, RRH, and antenna systems from the same vendor.

The definition of the Third Generation Partnership Project (3GPP) split RAN architecture that we illustrated in Chapter 3 greatly increased the options available to operators to deploy RAN in such a way as to adapt the parts of the RAN to be distributed at the cell sites and the parts of the RAN to be shared in a remote data center in such a way

Open RAN Explained: The New Era of Radio Networks, First Edition.
Jyrki T. J. Penttinen, Michele Zarri, and Dongwook Kim.
© 2024 John Wiley & Sons Ltd. Published 2024 by John Wiley & Sons Ltd.

as to provide sufficient resources (capacity) to the users of the system and sufficient capabilities (RAN key performance indicators, or KPIs) to support the required services. The 3GPP split RAN architecture breaks up the base station into four different elements: the centralized unit (CU), the distributed unit (DU), the radio unit (RU), and the antenna system (AS). Of these units, only the antenna system on given radio frequency (RF), has to be located at the cell site, while the CU, DU, and RU may be located either with the RF or remotely. 3GPP offers eight possible ways to disaggregate the base station spanning from option 1, where all the base station elements are co-located with the RF and therefore resulting in a fully self-contained bases station, to having all the base station elements remote and only the RF at the cell site (option 8). When most of the base station elements, and therefore most of the signal processing is performed at the cell site (corresponding typically to options 1–5), the functional split is referred to as "high-level split" (HLS), while when signal processing is moved to a data center away from the cell site the functional split is called "low-level split" (LLS). LLS corresponds to option 6 and above and the most prominent example is the O-RAN Alliance split 7-2x, a variant of 3GPP option 7 split.

The decision on how the base station is disaggregated, that is, which parts of it are at the cell site, which parts are centralized and shared among several cell sites besides having an impact on the performance of the RAN has an equally profound effect on the costs of the RAN overall. It is clear that from a cost point of view, it would be preferable to maximize the amount of the base station elements that are shared among different cell sites since this will decrease the footprint, energy consumption, and complexity of the equipment deployed at each cell site. However, such full centralization of the RAN may not always be feasible for at least two reasons:

- The physical distance between the data center hosting the base station functional units and the antenna system introduces an additional delay and negatively impacts the latency. For some classes of services such as ultra-reliable and low latency communications (URLLC) services typically used for industrial IoT, the additional latency is not acceptable.
- The bandwidth of the interface between the antenna system and the base station may become prohibitively high and expensive. For reference, Figure 8.1 shows the bit rate expected on the interface between the two parts of the base station depending on the functional split.

With Open RAN, operators will be able to adapt the distribution of the functionality of the base stations in such a way as to strike the best compromise between providing the performance needed in different scenarios and the overall RAN costs.

A further advantage of the disaggregation of the base station using Open RAN technology is that it will results in much greater openness. Distributed RAN, centralized RAN, and Cloud RAN that we saw in Chapter 3 are *de facto* closed systems with very limited possibility of using multiple vendors and many proprietary interfaces that vendors have developed to optimize their products and gaining an edge over the competitors. The split RAN architecture has indeed resulted in the standardization of some of the interfaces between the base station units, but is only thanks to the work of O-RAN Alliance and other organizations

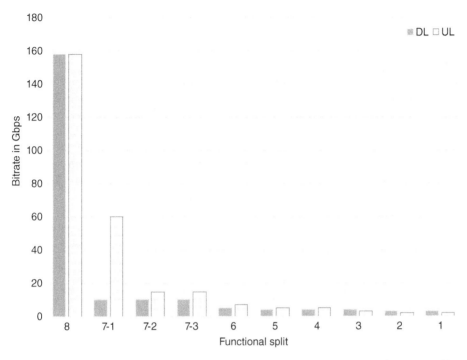

Figure 8.1 Bandwidth requirement for the connection between the units of a disaggregated base station depending on the selected functional split.

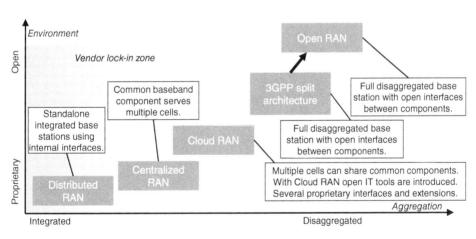

Figure 8.2 Mapping different types of RAN architecture against openness and level of integration.

such as the Small Cells Forum that we now have a fully disaggregated base station, with well specified, open interfaces. Leveraging the specifications of O-RAN Alliance and the ever-growing number of suppliers that support them, operators can now start experimenting with the deployment of different types of base stations where components from different suppliers can be integrated and interoperate successfully (Figure 8.2).

8.2   Analysis of the Functional Split of the Base Station and Performance

8.2.1   Need for Multiple Splits

The choice on the functional split of the base station results in a trade-off between cost and complexity. A low-level split, more advantageous in terms of equipment and running costs, requires a highly sophisticated transport technology and may not be capable of meeting some stringent RAN KPIs. On the other hand, a high-level split will not result in significant savings. As there is no one-size-fits-all, it is expected that there will be within the same network parts of the RAN that leverage the benefits of centralization to the full while other parts that will resemble a traditional distributed RAN architecture.

In the rest of this section, we will illustrate some functional splits that are, in the view of the authors, likely to be used in the field briefly discussing their advantages and disadvantages as well as the use cases and deployment scenarios they are most suitable for. These functional splits are shown in Figure 8.3.

8.2.2   High-Level Split – Option 2

In this configuration, the CU is centralized and serves multiple sites that host the DU and RU. This scenario does not pose particular problems in terms of transport between the data center and the cell site since the bandwidth of the signal after the DU processing is

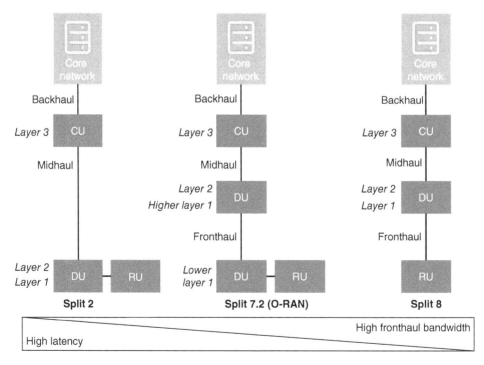

Figure 8.3   Comparison of different RAN splits.

quite manageable. This interface is at times referred to as midhaul to distinguish it from the backhaul that connects the base station with the core network and the fronthaul that is instead the interface between DU and RU.

While it may be argued that there are negligible advantages in this RAN architecture compared to the traditional architecture, it should be noted that the interface between CU and DU is fully specified in 3GPP and therefore it lends itself to a multi-vendor configuration. Furthermore, the fact that the midhaul specifications are not very demanding in terms of bandwidth and synchronization makes the deployment of a single CU possibly several tens of kilometers away from the cell sites it controls. The CU in this configuration controls many cells and can also be virtualized, optimizing hardware utilization. Another major advantage, though not specific to this option, is the implementation of the Control Plane – User Plane Separation that, as seen in Chapter 3, has already been introduced in the core network: the midhaul interface, called F1 in 3GPP, is split into F1-C that takes care of control plane messages and F1-U that transports the user plane data.

Option 2 is a suitable deployment choice for the case where coordination between different cells is not a required. Examples include fixed wireless access applications which deliver high bandwidth services to users in fixed locations. Another interesting application of Option 2 is where the CU controls groups of cells that are busy at different times of the day such as a residential district and a commercial district and the CU resources are used for different cells at different times.

Another example where Option 2 may be the only viable solution is where mmWave access points are deployed. As the bandwidth of a mmWave channel could be in the region of several hundreds of MHz and tens if not hundreds of antennas will be used, the requirements for the fronthaul that other functional splits introduce could become excessive.

8.2.3   O-RAN Alliance Split 7-2x

This option, which is based on one of the three variants of option 7 defined by 3GPP, was discussed extensively in Chapter 4 where we noted that the main benefits of this functional split.

In terms of deployment scenario, an RAN architecture implementing the 7-2x split is suitable for urban and suburban deployments where the length of the fronthaul interface is in the order of a few kilometers and where coordination between different cells is highly valuable. Also, by performing some of the signal processing at the antenna system, it becomes feasible to use a very large number of antennas (MIMO, i.e., multiple input multiple output systems) with a manageable fronthaul bandwidth.

8.2.4   Small Cell Split

The small cell split envisages the inclusion of the full physical layer at the cell site. The small cells forum extended the specification of this split which is based on 3GPP Option 6 by defining in more detail the interface between the small cell DU and the small cell RU. In the view of small cell forum, this will foster competition among suppliers of small cells and widen the ecosystem. This functional split benefits from reduced demand on the fronthaul (most of the complex and computational expensive work is carried out in the RU) as well as efficient use of the hardware thanks to the centralization of the CU.

8.2.5 Low-Level Split – Option 8

In this functional split, all the base station components, except for the RF, are co-located. 3GPP defined this option for completeness but agreed that no further specification work was needed. The main problem with this option is that the bandwidth of the transport interface between the RF and the RU could easily be in the order of hundreds of Gbps in case of 5G. The option is still relevant in two instances:

- For Legacy 2G and 3G radio access networks. These systems utilize a relatively narrow channels (e.g. 5 MHz in 3G compared to 100 MHz for 5G) and few antennas therefore the fronthaul bandwidth is within the capabilities of enhanced common public radio interface (eCPRI). The advantage for operators in using Option 8 is to leverage virtualization of base station components that are expected to be implementable on an off-the-shelf ×86 processor as well as to be able use different vendors on the interface between the site (which for 2G/3G comprises of antenna system and RRH) and the data center (where the rest of the base station system is located).
- For highly specialized 5G services that require ultralow latency, high-order MIMO, and extremely tight coordination between different cells. Typical use case could be a factory floor or a private network. The problem of the extremely high bandwidth on the connection between antenna system and the rest of the base station is mitigated by the small physical separation between the two parts.

8.3 Service-Based Planning Aspects

8.3.1 New 5G Services and Planning

The services supported by 5G services extend beyond the traditional broadband scenarios. The two additional groups of 5G services are URLLC and massive IoT (MioT). For example, the 5G network will need to support the new intelligent transportation services and mission critical communications with high carrier-grade performance. Furthermore, even the traditional broadband services are extended to high-resolution 8K videos and immersive media such as the augmented reality/virtual reality (AR/VR). This puts further pressure on Open RAN to match the performance of traditional RAN, where the requirements are becoming more stringent.

Further to the services considerations, there are various types of deployment scenarios that the Open RAN needs to address if it is to replace the traditional RAN. For example, it may be better to employ split 7.2 or higher if the RAN is being deployed in urban scenarios, where users will be more demanding and transport infrastructure will be more likely to be accommodating. This is not the end of the story, and the Open RAN deployment must also take into account legacy networks if it is brownfield deployment. Even in the case of greenfield deployment, there is no legacy network to fall back to, and considerate service planning is required.

In this context, this section provides considerations in planning the network for diversity of 5G services in various perspectives. First, the challenges of operators, including multi-vendor scenarios, performance, and cost pressure, will be discussed. Then, the challenges of Open RAN vendors will be covered with consideration of factors important in real life such as reputation and roadmap building. Finally, the considerations will be applied to brownfield and greenfield deployment scenarios, where challenges will be described and potential remedies given.

8.3.2 Challenge of Operators

8.3.2.1 General

Adopting Open RAN comes with potential benefits of flexibility and expanded ecosystem of network equipment, but it also comes with additional challenges that operators need to address. Multi-vendor scenarios add complexity in integration and may also limit or constrain the improvement in performance. In addition, both the cost structure and the skillset required in Open RAN may be different from that of the traditional RAN, requiring operators to adapt to the new paradigm and environment. Therefore, these challenges are worth investigating such that the operators planning to deploy Open RAN can identify the potential issues beforehand and prevent/mitigate them.

8.3.2.2 Multi-vendor Integration

Open RAN, by definition, leans toward multi-vendor scenario. It is indeed possible for a single vendor to be selected to deploy the RAN based on Open RAN, but the intention behind disaggregation and opening up interface is to allow various vendors to be selected to deploy the RAN. Therefore, Open RAN is very likely to be deployed in multi-vendor scenario, where integration can become a challenge for the operators.

Having more vendors can indeed lead to potential cost savings, and the utilization of standard COTS hardware can simplify the supply chain of RAN [1]. However, more vendors mean that there will be more stakeholders to deal with, and resolving an issue in more complicated stakeholder environment will be more time-consuming and challenging. Furthermore, splitting the previous single-vendor deployment to multi-vendor deployment means that each vendor will receive less than the single-vendor case, because the deployment cost needs to be split. This may discourage some vendors in investing resources to satisfy the complex operator requirements and/or to troubleshoot issues [1]. While clear ownership and accountability structure will remedy this issue, it is easier said than done and integrating the multiple vendors require experience and expertise in the operator side to be successful.

Multi-vendor scenario is also likely to accompany more issues in the beginning of the deployment, as the products are not from the same vendor and it will require fine-tuning and adjustments to ensure interoperability of the products. For example, open fronthaul comes with a variety of protocols in different planes as indicated in Chapter 4. This means that connecting and O-RU from one vendor to O-DU/O-CU from another vendor will likely to be not plug-and-play (i.e. they work upon connection) [2]. This requires technical coordination among vendors and possession of knowledgeable technological experts and integration engineers to remedy.

The complexity of multi-vendor integration is not limited to deployment, but also in operation of networks [2]. Especially in the context of service management and assurance, coordinating multiple vendors and integrating the system to achieve carrier-grade performance is more difficult than the case of having a single vendor. To remedy this, the operator needs to possess integration capability and/or appoint a system integrator that can coordinate the stakeholders and manage the complexity of the system. The latter, however, may come with a heavy dependence on the system integrator, which is contrary to the rationale behind Open RAN (i.e. to reduce lock-in).

8.3.2.3 Cost Challenges

The potential benefit of Open RAN is to optimize cost by disaggregating traditional architecture and opening interfaces. Theoretically, multi-vendor scenario will enable operators to select best components among vendors while reducing cost because the RAN is unbundled. However, as discussed in the previous Section 8.3.2.2, the multi-vendor scenario does not necessarily lead to cost savings because of the multi-vendor integration challenges. Furthermore, the total cost of ownership (TCO) depends on the topology of the RAN and also on the types of technologies that are employed (e.g. automation, cloud platform, and lower processing latency) [3]. It is true that the pooling of the baseband unit (O-DU and O-CU) resources in the edge data center can potentially enable significant const savings, but the scenario requires extensive fiber network availability in order to meet the latency requirements and also the edge data center availability. Since each of them is itself a challenge, the pooling of the baseband unit is not yet a readily available option for the operators [3].

The TCO consists of two aspects, capital expenditure (CAPEX) and operating expenses (OPEX). Depending on the context, Open RAN may come with varying degrees of both, and therefore it is necessary to examine the factors behind the two aspects. The first major factor behind CAPEX is investment in hardware. Even though Open RAN is based on whitebox standardized hardware, there are a variety of hardware that needs to be purchased. It is required to purchase the RRU for radio interface, and the switches and COTS server for aggregation and cloud setup [3]. As mentioned previously, the edge data center setup is required at least initially, and this will be a pressure to increase CAPEX. In the long-run, however, it is expected that the already deployed edge data centers will be available for utilization, and less investment will be required than was initially. Similarly, the upgrades to the existing transport network or deployment of fiber network will be necessary to accommodate Open RAN fronthaul requirements [3]. Initially, these costs will be significant and increase the CAPEX, but will reduce CAPEX in the long-run as the upgraded/deployed fiber networks will be available for use.

The second factor behind CAPEX is software investment. Unlike the traditional RAN, there is an additional virtualization layer whose software needs to be procured. That is, on top of the network functions such as CU and DU, hypervisors, host operating system, and complete stack of Openstack are required to deploy Open RAN [3]. Of course, the CAPEX would depend on the licensing model of the software, but licenses for procurement and maintenance are not necessarily bundled and having a separate maintenance license can further increase the CAPEX. Therefore, careful consideration on the licensing model is crucial to optimize the cost.

The last factor of investment in automation is related to the first and the second factor, but stands out in Open RAN deployment because it is one of the three key differentiator of Open RAN (i.e. intelligence). Because Open RAN introduces the service, management, and orchestration framework to employ automation technologies such as the machine learning, the hardware and software for these also need to be procured. While the investment on the automation will increase OPEX, the CAPEX will inevitably increase.

For OPEX, the consideration is more subtle as many of the factors are not easily quantifiable, meaning that experience and knowledge in the area will be crucial to realize savings. The first factor is operation and maintenance cost [3]. The change in the operational model of RAN in Open RAN can be a double-edged sword. The automation can reduce the OPEX, but operating the virtualized environment and split in the RAN protocols may increase the OPEX.

The second factor is the cost for testing. Recall from the discussions in multi-vendor integration that to integrate different vendors' components will require extensive testing and fine-tuning. Further complication comes from the fact that different products and software have different lifecycles, and as lifecycles become shorter for virtualized products, the required number and degree of test will ever increase. To optimize these costs, appropriate skillset and experience are required.

In the era of climate change, energy cost is an important consideration not just in terms of OPEX but also in carbon footprint. Open RAN comes with two opposite effects for energy cost. The introduction of general-purpose hardware such as COTS will increase the power consumption, as dedicated hardware tends to consume less power [3]. On the other hand, the centralization of hardware in cloud data centers can benefit from economies of scale in cooling, lighting, and electricity consumption, which may lead to less consumption and less energy cost [3]. Just as in the case of other OPEX costs, skillset and experience will need to be accumulated to optimize energy cost.

8.3.2.4 Performance

Dedicated hardware/product is likely to be more expensive than general-purpose hardware/product, but it is usually selected because of better performance. The operators face similar conundrum in Open RAN. While the general-purpose hardware/product can lead to potential cost savings and more flexibility, there is a trade-off between the benefit of Open RAN and performance. Indeed, many expect that the multi-vendor Open RAN network may not have the performance or security matching those of the traditional RAN [3]. The traditional proprietary RAN systems come embedded with vendor-specific know-how to increase performance, and advanced functionalities like the digital beamforming and MU-MIMO can be limited in Open RAN because this specific know-how is not present in the Open RAN system [1].

Nevertheless, the introduction of automation layer, that is, service management and orchestration (SMO) and radio intelligent controller (RIC) along with the centralized policy-driven user data analytics can enhance the performance of Open RAN, reducing the performance gap [3]. Furthermore, the loss of performance in general-purpose hardware can be offset with adoption of hardware accelerators such as field programmable gate array (FPGA) and application-specific integrated circuits (ASIC) [3].

8.3.2.5 Skillset and Experience

Acquiring relevant skillset and experience for Open RAN is probably the most important and the most difficult challenge that the operators face. As discussed in the other challenges, the skillset and experience regarding Open RAN can offset the disadvantages of Open RAN and maximize the advantages of Open RAN. Indeed, relevant skillset in operation and maintenance is crucial to minimize the associated cost and the disruptions caused by outages and issues. Integrating different vendors' component will be less challenging if the operator already possesses integration skills and experience.

The skillset and experience do not apply only to the case where the operator plays an integrator role, but also apply to the case where there is a separate system integrator. Even when the operator is merely supervising, relevant skills and experience are needed to identify potential issues and effectively coordinate stakeholders.

The challenges for operators in acquiring the necessary skillset and experience are that both hiring of new talent and retraining/upskilling of existing talent need to be progressed concurrently. Telecoms industry is a very specialized industry that differs from IT especially in terms of requirements, and therefore there is always lack of personnels who have the relevant background and domain knowledge. In this context, hiring of new talent is challenging because there is not enough people and retraining/upskilling of existing talent becomes crucial. Therefore, the operators need to implement comprehensive training for its existing workforce such that the workforce can acquire hands-on experience in every functional block or system component, as well as being able to interact with not only a pool of vendors but also diverse standards and open-source community to understand the implications of Open RAN [1].

8.3.3 Challenge of Vendors

The Open RAN opens up opportunities for vendors with a new paradigm of deployment and operation, but the reaping the fruit of the opportunities comes with challenges. In the vendor perspective, the multi-vendor scenario presents them with multitudes of settings and component sets that the vendor needs to support. This implies that the vendor not only needs to test against these configuration (putting aside the total number of possible configurations) but also needs to synch its product roadmap with the vendors of other components to ensure interoperability. While it is possible to add a degree of differentiation to its product, synching roadmap means that there will be constraints in developing a new feature and enhancing its product. Therefore, providing differentiation while ensuring interoperability will be a great challenge for Open RAN vendors.

Another major challenge is to respond to a variety of requirements and varying market demand. Unlike the traditional RAN where the RAN was procured at most to the granularity of regions, Open RAN enables operators to procure equipment flexibly. This means that it is difficult for vendors to make a stable forecast of demand for its products and scaling and descaling its manufacturing in line with the market demand can be difficult. Some markets are more demanding in terms of performance than others, and some are more demanding in terms of cost. Trying to match the performance of the traditional RAN in the former will be a challenge that the Open RAN vendors need to address.

Consequently, the Open RAN vendors will face increasing cost pressure. In addition to the factors mentioned earlier, the Open RAN vendors will need to tackle the challenge of recovering R&D investment and paying royalties for existing cellular communications patent pool. The patent system in cellular communications and R&D investment to build the patent pool is a significant investment, and the Open RAN vendors must be able to control this investment while building on top of the varying market demand.

However, this is not the end. The Open RAN vendors are likely to be nontraditional incumbent vendors and will likely to have less reference and/or business relationship with the operator customers. To be able to replace the traditional vendors, the Open RAN vendors need a track record and a kind of proof that they will provide benefits (i.e. cost/performance) compared to the traditional RAN vendors. On top of this, the Open RAN vendors need to demonstrate that they have the capability of cooperating with other vendors to upgrade and troubleshoot. These altogether are not built in a day and requires some time to be translated into reputation. The Open RAN vendors therefore need to be ready and willing to build reputation for long period of time.

8.3.4 Considerations in Brownfield and Greenfield Scenarios

8.3.4.1 Brownfield Deployment

Brownfield deployment refers to the case where there is an existing infrastructure and the deployment involves upgrading of or addition to the infrastructure. In brownfield deployment, the operators and vendors need to consider four aspects to successfully deploy Open RAN: integration of open interfaces and RIC, integration of O-RU, disaggregation and cloudification of network functions, and hardware performance. These are described in detail in [1], of which this section provides a high-level overview.

The integration of open interfaces and RIC in the legacy network requires the operators to upgrade already deployed long term evolution (LTE) macro eNBs to support the X2 C-plane and U-plane interfaces with O-CU-CP and O-CU-UP, respectively. Since the brownfield operator would be relying on its legacy 4G LTE network for major portion of subscribers, it may be difficult to make such upgrades and ensure quality/performance (e.g. by testing) of key procedures such as load balancing and dual connectivity when the new entity O-CU (CP and UP) is added. Similarly, the LTE macro eNBs need to support the E2 interface between the eNBs and near-RT RIC. This means that the splitting of radio resource management can be challenging in the same context because new entity needs to be supported by legacy network nodes.

In terms of O-RU integration, recall that O-RAN Alliance adopted Option 7-2x split between the O-RU and O-DU, which requires the transport network to support a degree of bandwidth. The transport network supporting legacy LTE, however, may not be adequate to support this split and the integration of O-RU in this scenario can be challenging. In addition, some of the 5G features supported by the O-RU and O-DU may require different split/architecture to be optimally supported in the legacy 4G LTE network. Adopting multi-vendor paradigm in O-RU can also be challenging, where different physical layer implementation in O-RU by different vendors can limit features such as interband carrier aggregation.

The brownfield operators have challenges of disaggregating the existing RAN network functions and to virtualize them so that they can be migrated to cloud platform as cloud network functions. The resource dimensioning needs to be done carefully, as the required computational resources and storage spaces in the cloud platforms can be different from the legacy network hardware resources. The management aspects of virtualized cloud platform also need to be considered. The distribution of existing network functions for 4G LTE networks is likely to be dispersed, and some of the functions will need to be centralized in data centers and some will need to remain. Since this is in the context of virtualized cloud platform, the migration of functions will not necessarily be one to one between legacy site and data center. Therefore, the planning of the resource distribution needs to be done carefully. Furthermore, the user plane load requirement of mobile networks makes selection of cloud provider and cloud infrastructure resources challenging, as there are not many providers that can meet the requirements.

The advances in accelerators and generic hardware performance can minimize the gap between the performance between traditional RAN and Open RAN as described in Section 8.3.2.4. Nevertheless, there are additional hardware performance considerations that brownfield operators need to take, such as validating the end-to-end deployment consisting of multitudes of protocol stacks and technologies of both legacy and Open RAN technologies. Furthermore, ensuring power management/distribution of the servers is critical to ensure reliable operation and services of the RAN equipment in data centers.

8.3.4.2 Greenfield Deployment

Greenfield deployment refers to the case where there is no existing infrastructure and the deployment is done with a clean slate. In greenfield deployment, the operators and vendors need to consider four aspects to successfully deploy Open RAN: evolution of Open RAN interfaces, realization of Open RAN network functions, integration of disaggregated network functions, and cloud footprint and scaling. These are described in detail in [1], of which this section provides a high-level overview.

The greenfield operators do not have the restrictions of legacy interoperability of brownfield operators, but they face the challenge of implementing the Open RAN specifications that may not be fully mature or ready for their market. This means that the specifications may evolve, and the current deployment may need to be upgraded significantly in the future. Furthermore, as Open RAN specifications employ cloud native technologies, it may not be fully compatible with 3GPP specification-based RAN solutions. For example, operation and maintenance features of Open RAN may not be fully compatible with those of 3GPP specification-based RAN solutions and may require further integration efforts.

In a related manner, the realization of the Open RAN network functions in practice poses challenges in meeting the performance requirements. For example, the cloudification of O-DU and centralization of the baseband unit means that there will be more challenges in meeting low latency requirements than traditional RAN, where the physical cell sites are much closer to the radio units. This has implications in key RAN features such as the modulation and coding scheme, where synchronization and real-time granularity are critical in ensuring quality.

Security is another issue associated with disaggregated Open RAN network functions. The existing security and trust models are not applicable to the disaggregation model, as traditional model involves a number of physical equipment that are owned/operated by trusted party (usually the operator itself). With cloudification of RAN functions, the RAN functions are likely to be co-located with those of other operators/players and sometimes with other non-telecom tenants of cloud services. The security process therefore needs to be enhanced from the traditional process, but this is a new experience for the telecom industry, and it may take some time and field experience to mature.

The Open RAN deployments come with RIC functions that utilize AI/ML technologies. The training of the AI/ML models requires huge computation and storage, and this is not just one-time resource consumption but continuous because the AI/ML models need to be updated based on the accumulated network data. This indicates that the optimization benefit by RIC can be diluted by the high consumption of cloud resources and increasing demand for these resources as network grows (i.e. having more UE and network nodes will exponentially increase the number of transactions per second). The greenfield operators relying completely on Open RAN must be mindful of these potential cost factors to maximize the benefits.

8.4 Testing and Measurements

8.4.1 General

Although the O-RAN concept aims to provide fluent and seamless interoperability between the involved stakeholders, including vendors of several network components as well as roaming operators, in practice, the O-RAN disaggregation is complicated considering the complete set of protocol layers. Figure 8.4 outlines some of these challenges as interpreted from [4].

As an example, the O-RU needs to process time-sensitive functions in real time, which sets highly demanding requirements for the lower physical layer components. At the same time, the components of the O-DU need to be able to handle the signaling under unpredictable and changing radio channel conditions including fading scenarios due to UE mobility and respective varying radio link attenuation because of buildings, vegetation, vehicles, and other obstacles occurring between transmitting and receiving antennas. The upper-layer protocols are also impacted due to the changing capacity demand due to the varying number of users and traffic volume. At the application layer, some applications are especially demanding for the quality of the radio channel and system processing such as those based on virtual and augmented reality, and the underlaying infrastructure needs to be adjusted, respectively.

The correct dimensioning of the network – both radio and core network segments – is thus critical for the fluent user experiences. The deployed network is never static in real life as the utilization profiles change, and network components may have failures in the course of time. This means that constant testing and measurements are needed to ensure the correct quality level of the network, as well as the expected functioning in all the scenarios – including the interoperability, which according to Spirent survey with Light Reading [5] is the considered as the most challenging single item in the Open RAN deployment, the actual performance aspects being clearly less worrying together with new security, OAM, and system integration overhead.

Performance-wise, not only the operational network phase requires constant quality monitoring, but the initial setup needs to be adjusted correctly involving conformance and interoperability testing. Any later changes in the deployed network, such as upgraded

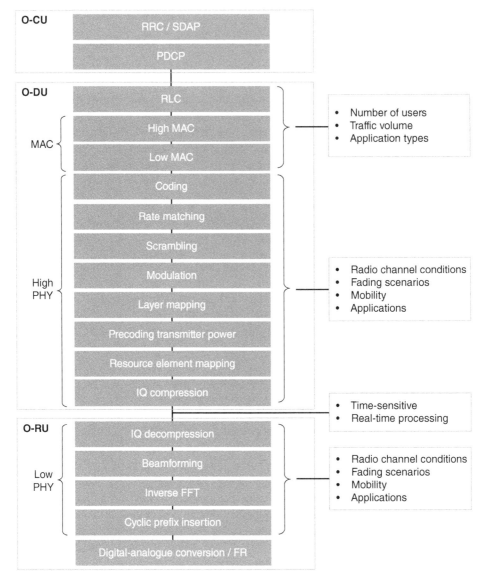

Figure 8.4 Challenges at O-RAN protocol layers requiring testing, as interpreted from the O-RAN testing guidelines of Spirent. Source: Adapted from Spirent [4].

software or hardware during the lifetime of the network, may also require thorough testing for ensuring the correct functioning of the interoperability aspects.

As indicated in the research conducted by Spirent, and depicted in Figure 8.5, the most important challenges the ecosystem sees at present relate to the correct solving of multi-vendor interoperability, as well as adequate performance and robustness. [4] The up-to-date measurements are needed to overcome these challenges.

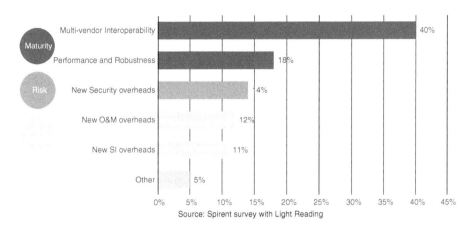

Source: Spirent survey with Light Reading

Figure 8.5 The most relevant challenges the ecosystem sees in Open RAN deployment. Figure by courtesy of Spirent.

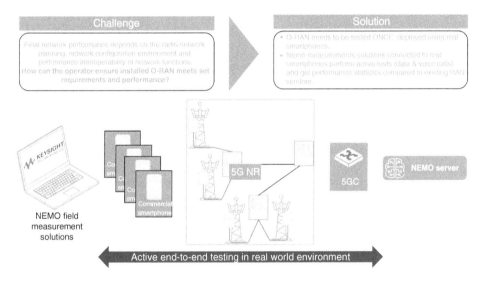

Figure 8.6 Challenges and solutions related to Open RAN deployment. Source: Figure by courtesy of KeySight Technologies.

Among other test equipment ecosystem representatives, KeySight indicates typical challenges in Open RAN deployment as depicted in Figure 8.6. The end-to-end testing is beneficial for the verification of the Open RAN deployment. Upon the network setup, the realistic environment can be evaluated in various means, including smartphone-based solutions that suit for voice and data service testing. The results can be assessed comparing the performance figures of the existing RAN and the new Open RAN network.

Regardless of the thorough standardization of the Open RAN interfaces, in practical networks, there will always occur occasional faults for which fault management (FM) is needed to complement the PM to capture the faults (alarms), prioritize the fault types, and to provide process for the reparation and recovery within the corresponding time window for the criticality of the fault. The adequate process for fault recovery on time is of utmost importance for maintaining the agreed service level assurance (SLA). In the worst case, the failure to comply with the SLA may result in additional costs and could cause a chain reaction between the parties having the SLA in place. Furthermore, failed SLA level ultimately impacts the end-users' perception of the quality that may damage the reputation of the Open RAN stakeholders.

The constant PM and FM, complemented by high-quality configuration management (CM), can largely prevent the issues before they even materialize. As an example, some less critical alarms may indicate the wearing out of some of the components, and timely reaction to the issue can save lot of later troubles if the component can be replaced before it fails completely. Open RAN may generate completely new type of alarms, so the operators need sufficient time to start understanding the practical significance of respective alarms.

As commented in [6], while the transition to Open RAN offers benefits for MNOs, it also generates additional complexity for the testing of the compliance with the requirements.

8.4.2 New Methods and Challenges in Open RAN Testing

Many of the needed testing methods for performance monitoring are like the ones applied in any mobile communication system prior to the Open RAN concept. Nevertheless, the new technical solutions related to Open RAN require also new methods, as indicated, e.g. in [5]. These methods include wrap-around testing and emulation of the radio access network components, interfaces, and RF channels relevant to the Open RAN. Along with the advanced functioning of Open RAN, such as that of near-real-time RIC, automated means are beneficial and needed to adequately assess the quality. The testing that considers Open RAN-specific aspects is beneficial for ensuring multi-vendor interoperability and for achieving the expected KPIs. The end-to-end chain of the Open RAN testing environment includes the actual Open RAN components as well as the UEs and core network connecting to it. Figure 8.7 depicts this landscape [4].

As indicated in [6], the three key measurement categories for Open RAN are the conformance testing, interoperability testing, and end-to-end (E2E) testing.

Conformance testing requires detailed test scenarios and the compliance verification of the Open RAN device under testing (DUT), the device being, e.g. O-RU or O-DU. For this, the functions and interfaces of the Open RAN need to be implemented carefully ensuring the compliance to the 3GPP and O-RAN Alliance conformance test specifications and their related test case descriptions.

Interoperability testing refers to the interoperability of Open RAN DUTs. Examples of these components are the O-RU and O-DU. The functions and interfaces relevant to the interoperability must be implemented according to the O-RAN interoperability test specifications and respective test case descriptions.

End-to-end testing considers testing of a complete network that involves multiple Open RAN vendor components. The test cases need to comply with the 3GPP RAN setup. In this test type, the O-RAN network functions are considered as the DUTs. In the E2E testing, both 3GPP and O-RAN test specifications are applied.

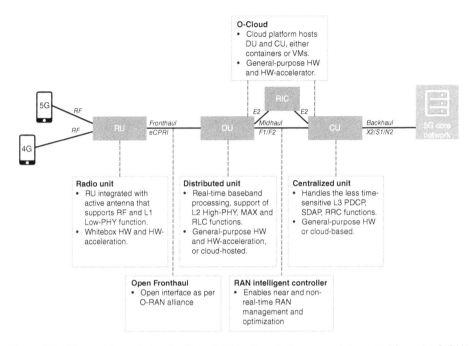

Figure 8.7 The end-to-end view for Open RAN testing challenges, as interpreted from the O-RAN testing guidelines of Spirent. Source: Adapted from Spirent [4].

Table 8.1 summarizes the key test specifications related to these three types of the Open RAN testing as indicated in [6], and Figure 8.7 summarizes the key areas the testing of the Open RAN need to consider as indicated in [4].

As explained in [5], the Open RAN concept results in the following new aspects involving new type of measurements:

- *Multi-vendor interoperability*: being still rather novelty concept in practice, the open "plug-and-play" environment requires new testing methods that considers interworking and the compliance of the correct functioning of open interfaces.

Table 8.1 Test specification references for Open RAN test types.

Conformance testing	Interoperability testing	E2E testing
O-RAN Alliance WG4 conformance test specification for M-plane, S-plane, and C/U planes including FR1 FDD/TDD and FR2 TDD.	O-RAN Alliance interoperability test specification for M-plane, S-plane, and C/U-planes.	O-RAN Alliance TIFG E2E testing specification for functional, performance, services, security, load/stress, and RIC-enabled E2E use case testing.
3GPP RF conformance test specifications for transmitter and receiver.		Additional test cases may consider 3GPP features, energy efficiency, performance, scalability, security, stability, robustness, resiliency, and OAM aspects.

- *Economic compliance*: one of the expected benefits of Open RAN is the cost-efficient RAN solutions that provide technical and economic benefits over the "traditional" radio access networks. Nevertheless, there may be pressures for ensuring the adequate cost level due to additional test and integration efforts that are required for implementing and operating the Open RAN environment complying the SLA.
- *Performance monitoring*: radio access is highly sensitive to quality changes and the end-to-end network performance depends largely on it. It is expected that Open RAN can maintain, or preferably improve, the radio performance and thus the experienced end-to-end quality level. Enhanced network performance measurements and complementing FM and other network performance-related methods are thus needed to ensure the expectations are met.

Furthermore, as stated in [4], the Open RAN assessment may include various items, such as standards conformance, multi-vendor interoperability, feature conformance, performance and robustness, mobility scenarios and handovers, security, capacity and scalability, synchronization, and timing, as well as application and service validation.

Along with the differences in Open RAN, the optimal deployment requires new testing methods and solutions for the novelty parts whereas the common testing with the "traditional" networks still applies. Being still new, Open RAN benefits from recommended best practices for validating its correct functioning and adequate performance. As is the case for the actual optimal functioning of the Open RAN system, also the measurements benefit from their automatization. In the initial phase of the network implementation, it might not be possible to test final commercial solutions, so network emulation and traffic generation can be applied to model the real world.

In addition, the precommercial silicon planning and development has a very important role in the efforts to achieve adequate performance figures upon network deployment and operation.

Adequate testing during the silicon development, including prototyping and timely chip-level adjustments, is thus important as the compromises done at this level may result in long-lasting issues in the actual operations that only are possible to correct by replacing critical equipment.

8.4.3 Conformance and Interoperability Testing

8.4.3.1 General

Open RAN constructs upon standardized open network interfaces as per the relevant technical specifications for radio of the 3GPP, O-RAN Alliance, ITU-T, and IEEE. The node functions need to meet the requirements of the 3GPP and O-RAN Alliance for the interfaces involving Open RAN components, and validation is thus needed for the adequate compliance under select Open RAN synchronization configurations valid for the implementation of interest.

Although Open RAN is assumed to ensure a facilitator for fluent interconnectivity of various vendors, there can be assumed to occur discrepancies as for with capabilities and parameters for configuration, e.g. in the O-RAN YANG data models of the devices [2]. The issues may be related to inconsistent parameter values that result in incompatible values or wrong types between devices, missing MNO-required parameters, additional or invalid parameters conflicting MNO requirements, and wrong character string leading to conflicts.

Vendors' test labs can be used to validate individual node functions considering adjacent network functions of the same and various other vendors for seamless interelement working. The testing is then extended to cover also end-to-end network aspects involving the services of interest such as voice, data, A/V, and emergency calls. The testing of the distributed architecture scenarios of Open RAN involving multiple vendors requires rather complex setup. The scenarios may include hosted data center or standalone cloud in physical or virtual environment and different security models.

Regardless of the aim of Open RAN to support open interfaces much more fluently and effortlessly than in previous radio networks, adequate system integration involving multi-vendor equipment and software is important in disaggregated and virtualized RAN. Skillful personnel are in key position to adjust correctly and optimally functioning multi-vendor environment. In this landscape, the stakeholders need to agree upon the share of the effort and liability; as an example, it is important to clarify, which party is the lead integrator (single entity or a combination of involved vendors, considering existing and new companies, or a separate entity). Also, along with the setting up the new network, it is important to understand the cost impact of ensuring the open interfaces versus maintaining the original, closed interfaces as a function of time, and whether the new architecture might lead into any closed ecosystem parts by the new vendors.

An important aspect is also the forming of adequate roaming and interconnection strategy between different access network types within and between operators and offload strategies considering foreseen scenarios.

8.4.3.2 Case Study: DoCoMo

As an example of the real-world implementations of the O-RU and O-DU integration challenges [7], NTT DOCOMO describes a scenario where vendors are required to satisfy interoperability testing (IOT) profile in early stage prior to the actual implementation to ensure adequate testing later with the MNO. In this landscape, some vendors offer O-RU as well as O-CU and O-DU solutions, making them capable of performing such as preliminary single-vendor IOT testing prior to the delivery of the solutions to an MNO. Nevertheless, single solution providers that have only O-RU or O-DU/O-CU can perform the initial testing through emulators for O-RU or O-DU functions. It can be assumed that such emulators are able to replicate sufficiently well the specified features for ensuring sufficiently good level of intervendor performance.

Another option for multi-vendor testing and certification is the use of open labs, such as those of OTIC and TIP Community. The OTIC-recommended testing procedure is depicted in Figure 8.8 for optimal integration efforts [8].

Figure 8.8 presents an example on how the vendors and MNOs can reduce the integration time and cost. The preliminary testing as soon as feasible already prior to the commercial network implementation facilitate the troubleshooting and minimizes the time to market as the issues can be solved in early stage. Nevertheless, as the integration of O-RU and O-DU can be rather time consuming in practice, to optimize this effort further, vendor can use automation through predefined script. It works as a feasible and cost-efficient way for testing the interoperability of relevant components, whether

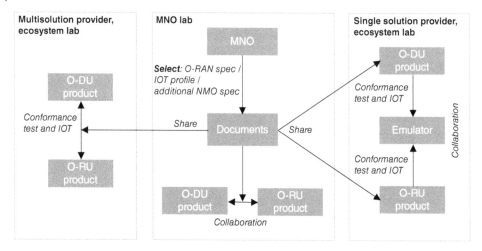

Figure 8.8 Interoperability testing procedure for the O-RU and O-DU. Source: Adapted from [8].

commercial or emulated, including UE, O-RU, and O-DU, and considering practical Open RAN parameter settings.

As an example, DoCoMo has informed about their efforts to develop an automated test script suite in order to reduce the O-RU test and integration time [7].

8.4.3.3 Case Study: Vodafone

Also, Vodafone is cooperating with test solution providers to implement O-RAN test solution. Furthermore, Vodafone has produced a guideline for Open RAN testing [6]. As an example, Vodafone divides the Open RAN test labs into three categories that are focused on the following aspects:

- Functional end-to-end testing regarding 4G eNB and 5G gNB that can be conducted through basic lab setup, with test objective for Open RAN technology exploration, assessment, and evaluation. This lab category can be based on the end-to-end test cases provided by the O-RAN Alliance Test and Integration Focus Group (TIFG).
- A step more advanced test labs, requiring more elaborate setup and knowledge, can study the level of 3GPP conformance, O-RAN conformance, and O-RAN interoperability. This category is suitable for detailed testing of massive MIMO beamforming radios, O-RAN DUs, and base stations, and the conformance test specifications can be based on the documentations of 3GPP (base station conformance), O-RAN WG4 (interoperability tests), and WG5 (F1, X2, Xn).
- The third category is the most advanced. The focus of these labs can include testing and optimizing for features, energy efficiency, performance, scalability, security, stability, robustness, and resiliency, considering realistic operating conditions. The testing can be based on the O-RAN WG2 (nonreal-time RIC and A1 interface), WG3 (near-real-time RIC and E2 interface), WG8 (open reference stack), WG9 (xHaul transport), and WG11 (security) specifications.

Vodafone has developed a step-by-step testing methodology help the ecosystem to explore and test the key aspects of the Open RAN [6]. The related activities include the definition of test configuration with participating vendors, definition of the test plan, execution of the test cases, verification of the results, and sharing of the test report. Reference [6] emphasizes the importance of the test automation for the network functions verification under large number of scenarios prior to the actual network deployment, to minimize the later issues on commercial environment. For this, reference [6] presents concrete solutions and architectural models for test automation.

8.4.3.4 Recommendations from the Field

Based on the case examples resented earlier, it becomes clear that automatic end-to-end testing provides a feasible way to reduce required integration time for Open RAN deployments and is thus recommended. Being skillful in this domain, vendors are encouraged to develop such test scripts, whether applied in common ecosystem labs or individual MNO test lab, considering MNO requirements and feedback.

8.4.4 Test Laboratories

Open RAN Alliance coordinates the efforts for facilitating Open Testing and Integration Centres (OTIC) [9]. These setups provide an open, collaborative, vendor independent, and impartial working environment to support the progress of the O-RAN industry ecosystem. The activities of the OTICs include:

- Awarding O-RAN certificates and badges.
- Hosting O-RAN Global PlugFests, which are periodic events that O-RAN Alliance organizes and co-sponsors to enable O-RAN ecosystem progress through testing and integration. In these events, vendors and providers join to test, evaluate, and verify their products and solutions.
- Conformance, interoperability, and end-to-end testing of O-RAN products and solutions.
- Providing reference setups and testbeds.
- Demos, community events, or trials to support wide adoption of O-RAN specifications and promote the openness of the O-RAN ecosystem.
- Workshops or tutorials to foster and develop technical skills of the O-RAN community.

The whitepaper "Criteria and Guidelines for OTIC" describes more detailed OTIC application and qualification process. Approved OTICs can issue certificates and badges in the O-RAN Certification and Badging Program. Furthermore, O-RAN Alliance's whitepaper "Overview of Open Testing and Integration Centre (OTIC) and O-RAN Certification and Badging Program" (April 2023) states that the O-RAN certification and badging program represents a comprehensive mechanism to ensure confidence in O-RAN-based products and solutions for both the operator and vendor communities [10].

The O-RAN certification and badging program is to minimize repetition of fundamental and common tests related to O-RAN products and solutions prior to their deployment. This, in turn, helps reduce MNO testing, promoting at the same time O-RAN products and solutions of vendors. As it gathers vendors, important aspect is also the possibility to test the level of interoperability among participating vendors.

It should be noted that the O-RAN certification and badging program is not meant to substitute operators' Open RAN products and solutions testing as those are needed prior to deployment. Also, an O-RAN certificate and/or badge is merely an indication of improved environment and confidence level for the selected O-RAN setup for reduced complexity, but it does not guarantee the passing of test cases. Further testing for ensuring fluent product and solution deployment is thus needed. The aim of the O-RAN certification and badging is to define well enough processes, procedures, templates, and data format for efficient execution of the tests and for sharing the results, referring to specific O-RAN test specifications that O-RAN Work Groups and Focus Groups have produced. The aspects include conformance certification, interoperability test badging, and end-to-end badging, based on O-RAN test descriptions:

- The O-RAN certification verifies that a vendor's product is compliant to O-RAN specifications. The O-RAN certificate relates to a single or multiple interfaces of O-RAN.
- The O-RAN interoperability (IOT) badging proves interoperability of the products that connect through an O-RAN interface or O-RAN-profiled 3GPP interface.
- The O-RAN end-to-end (E2E) badging validates the minimum requirements compliance for an end-to-end system or subsystem, including security aspects and other mandatory end-to-end features.

The O-RAN has defined certificates and badges for the following products and interfaces:

- O-RAN certificates for O-RU with Open Fronthaul interface, for O-DU, and for a combined O-DU and O-CU with Open Fronthaul interface.
- O-RAN IOT badges for O-RU and O-DU or combined O-DU and O-CU that are connected via Open Fronthaul interface; eNB and en-gNB connected via X2 interface; o gNB-DU and gNB-CU connected via F1-C interface; and two gNBs connected via Xn-C interface.
- O-RAN E2E badges for O-RU, O-DU, and O-CU, or their combinations that are included in the end-to-end system or subsystem.

The current list of OTCs, as per October 2023, includes 11 premises that operate in Europe, Asia and Pacific, and North America regions.

8.4.5   Testing and Open RAN Multi-vendor Integration

8.4.5.1   General
The following sections summarize some of the key impact areas in the Open RAN multi-vendor integration and operations. These aspects forming the Open RAN system cover Open Fronthaul, SMO, and HW/SW disaggregation as depicted in the test guidelines of Spirent [4]. These scenarios can include the following scenarios:

- *Open Fronthaul*: different vendors for BBU and O-RU. The O-RAN multi-vendor integration can consider the following components and protocol layers between the RU and BBU: (i) eCPRI (control and user plane, C/U plane); (ii) network configuration (management plane, M-plane); and (iii) PTP (synchronization plane, S-plane).
- *SMO*: OSS transformation.

- *HW/SW disaggregation*: O-Cloud HW/SW; O-RAN applications could be from different vendors.

The protocol layers of the Fronthaul include:

- *C/U plane*: Layer 1 Ethernet, Layer 2 Ethernet and VLAN, upper layer IP, UDP, and eCPRI. Please note that the IP and UDP are optional.
- *M-Plane*: Layer 1 Ethernet, Layer 2 Ethernet and VLAN, and upper layers' IP, TCP, SSH, and network configuration.
- *S-plane*: Layer 1 Ethernet, Layer 2 Ethernet, and PTP or SyncE.

Some of the aspects related to the services within the initial CPRI fronthaul include:

- In the initial phase, the BBU and O-RU are not capable of performing ad "plug and play."
- Initial O-RU contains more functions than later, resulting in a number of fronthaul-related parameters.
- Extended coordination required between vendors.

In this stage, the key tasks of RU include A/D conversion, up/down conversion, and power amplification.

To cope with the transition to eCPRI, some of the key mitigation aspects include:

- O-RU components require additional knowledge from the technical personnel.
- Interoperability testing requires augmented knowledge.
- Updated deployment process needed.

In this stage, the key functions of the O-RU cover also fast Fourier transformation (FFT) and cyclic prefix (CP).

8.4.5.2 Open RU Testing

Open RU testing requires performance, parametric, and system tests. Isolated wraparound testing is beneficial for the performance testing, so that the O-RAN infrastructure can be largely emulated, whereas the DUT can be located within the chain of components. The RU being the DUT, the physical radio network interface can be emulated including UE and modeled RF channel, to which the DUT is connected. At the other end of the DUT, the 5G core network as well as CU and DU can be emulated.

For the parametric testing, RU being the DUT, the 5G physical layer is possible to test independently in real time so that the DU providers use real-time logging to it and mark time accordingly the related stamps.

The systems test can include, e.g., CU/DU and RU in a form of system under testing (SUT), relying on emulated 5G core and emulated UE at its both sides.

In the RU testing, the wraparound testing principle, complemented by automation, helps achieve interoperability between vendors in a cost-efficient and fast way. The test components can be virtualized and reused.

8.4.5.3 Open DU Testing

The Open DU testing involves performance and system tests. Wraparound testing can be applied to the performance testing. The performance testing considers the functional validation of the open fronthaul and F1 interface, as well as the gNB-DU multi-vendor and gNB CP/UP performance.

For the system testing, integration testing can be applied to evaluate the QoE and QoS, the readiness of gNB (CU and UP) and user applications, as well as the transport design.

8.4.5.4 Open CU Testing

The CU testing involves performance and system tests. The performance tests can include, through wraparound principle, the functionality of F1, Xn, N2, and N3, as well as gNB (CU) scalability compared to the gNB-DU. It also evaluates gNB (CU) control and user plane performance.

For the system tests, integration tests can be conducted considering QoE and QoS, 5G readiness for gNB (CU) scale, and transport design for user plane scale. This testing type allows incremental testing and gradual experimenting of RAN elements. It is also aligned with cloud-native approach to services.

8.4.5.5 Open RIC Testing

The wraparound tests can cover also the Open RAN RIC, having it as DUT between the emulated CU/DU and the emulated counterparty of the RIC (that is, using Non-RT or Near-RT RIC as DUT for emulated Near-RT RIC and Non-RT RIC, respectively).

The tests can cover the correct functions of E2, A1, and O1 and can cover RIC use cases and performance as well as xAPPs. This test type is aligned with the cloud-native services.

8.4.6 Other Measurement Considerations

The measurement of Open RAN-specific items, as well as vendor-specific solutions, such as the AI/ML algorithms, may give valuable information for the operators in their efforts to optimize the performance of different environments. An example of such measurements may include the evaluation of the quality of the training of different vendors' algorithms, and their impact on the achievable end-to-end radio network performance. Being rather novel, the measurement of such Open RAN items requires still further development, though.

8.5 Optimization

8.5.1 Traditional Optimization and New Technologies

RAN is a complex system that requires carrier-grade performance, and this means that optimization is critical in meeting the requirement in the midst of changing environment. The traditional RAN comes with various techniques and technologies to tackle this aspect [11]. First, it is possible to optimize the RAN from the deployment phase. Depending on how the network is planned, the RAN performance can be optimized. For example, it can be possible to predict the development in a given region to equip the RAN to accommodate potential new residents/customers in the region. It is also possible to optimize the RAN if the devices used in the area are consumer devices or machines that require simple data connectivity. The optimization is critical in ensuring RAN performance during the operational phase [11]. The RAN performance is monitored with KPIs collected from field testing and network statistics. Then, RAN parameters are fine-tuned, and equipment is fixed/upgraded if necessary to ensure that RAN performance meets the desired QoS. For example, the cells with highest dropped call rate can be adjusted to reduce more dropped calls with the same effort.

The concept of self-optimizing network (SON) was introduced in LTE [11], where the network employs automatic optimization to dynamically respond to changing load, faults, and the circumstances of the network. The main focus of SON is to minimize the work of the technical personnel, such that the person can focus on more critical and value-added aspects of network optimization. Building on the works of SON in LTE, the traditional RAN has developed the following capabilities:

- Automated installation and commissioning
- Power optimization and saving
- Optimization of maintenance efforts
- Optimization of coverage and capacity
- Optimization of handovers
- Optimization of signaling

Despite the rich capabilities SON enabled, the traditional RAN was limited in that it was largely based on heuristics [1], sometimes limiting the intricacies in real-life deployments raised by the interaction of multiple RAN parameters may not be effectively resolved. The traditional RAN also relied on vendor's proprietary radio resource management (RRM) functionalities. Even though they are based on the 3GPP standards, the implementation of the standards can be different. In this context, the intelligence differentiator of Open RAN can resolve the two challenges. First, the AI/ML-based solutions of Open RAN based on the big data behind the RAN can explicate the intricate interactions of RAN parameters, making the optimization more dynamic and effective. Second, the Open RAN adopts non-RT and near-RT RICs that uses open interfaces and allows third-party applications to run on top. This decouples the intelligence from vendor proprietary solutions and ensure interoperability among solutions on top of facilitating more innovation in the RAN with participation of more vendors.

8.5.2 AI/ML in Open RAN

It is indeed correct that traditional RAN is also improving and developing with the advanced AI/ML technologies, and this per se is not the differentiator of Open RAN. However, the openness of Open RAN can create synergy with AI/ML technologies and can provide the following benefits [1]:

- Open interfaces facilitate data collection by harmonizing the data model and makes changes in configuration of such task easier
- Open APIs enable various algorithms and applications from in-house and third party to be tested and implemented
- Allows network operators to control and innovate at the pace they wish

- O-RAN defined optimization use cases[1] already address the key issues of RAN optimization (e.g. energy savings, mobility management, load balancing, and traffic steering)

The near-RT RIC supports AI/ML technologies in various ways. First, it provides the data pipeline, which collects the RAN data over E2 interface and provides the data for xApps over near-RT RIC to consume. The data can be used to train the AI/ML models of xApps such that the model can be adapted to the current circumstances and not out-of-date data. In addition, open API is defined for xApps to communicate with the near-RT RIC, as to enable interoperability of xApps and near-RT RIC from different vendors. The API can be used to train AI/ML models and/or to deploy updated AI/ML models. The near-RT RIC also supports Y1 interface, which is used to expose RAN analytics information to O-RAN internal and external functions. The consumers of the interface can subscribe to or request the RAN analytic information.

The non-RT RIC trains and updates the AI/ML models and provides guidance on policy for near-RT RIC to follow. In terms of the former, the non-RT RIC trains the AI/ML models, allows the consumers of the model (e.g. near-RT RIC) to store and retrieve the trained AI/ML models, and monitors the performance of the deployed AI/ML models for further optimization. The latter does not directly impact AI/ML models, but provides a direction (or intent) for the near-RT RIC in performing relevant tasks.

These altogether enable various optimization use cases including handover optimization, network slicing optimization, and automated optimization as described in Chapter 6. The O-RAN Alliance [12] highlights four major use cases of RIC that utilize AI/ML models:

- Predictive maintenance: predict equipment failures based on historical data.
- Dynamic spectrum allocation: adjust spectrum allocation dynamically based on real-time network conditions.
- Automated network optimization: identify efficient network configurations based on the large volume of network data.
- Energy efficiency: dynamically adjust the power of network elements depending on the real-time network traffic.

8.5.3 Open RAN and Evolution Toward AI-native Network

The Open RAN utilizes AI/ML technologies to provide advanced intelligence and optimization techniques, but this is not the end game. As highlighted by O-RAN alliance [12], the O-RAN RIC is adopted as the compromise between evolutionary approach of introducing the features in existing standards and revolutionary approach of specifying AI-based network from scratch (see Figure 8.9).

The evolutionary approach is where AI/ML technologies are built on the 3GPP standards in backward compatible manner. This is the best approach given consistency and compatibility with existing technology, but has a drawback in that it may delay innovation as development of standards can sometimes be time consuming. Furthermore, interoperability itself can sometimes become a constraint. If a technology has to be compatible with legacy technologies from far in the past, the technology cannot be as flexible as it would

---

1 See Chapter 6 for details

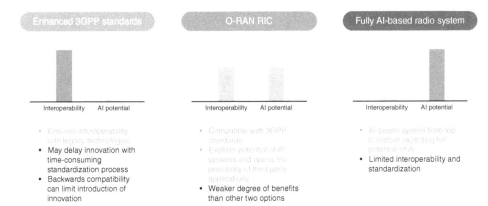

Figure 8.9 Comparison of approaches to AI-native network. (Bar graphs for illustrative purposes, not intended to provide absolute comparison). Source: Adapted from Choi [12].

be when it does not have to be compatible at all. This is precisely the reason why the 5G standards were (at least in 3GPP Release 15) developed without the requirement of interoperability with 2G/3G networks.

The revolutionary approach, on the other hand, is where the network overall is rearchitected with exploiting AI/ML technologies in mind, meaning that all network layers and building blocks of RAN specifications are built from scratch. While this approach maximizes the benefits and the potential of AI/ML technologies in RAN, the lack of standardization leads to limited interoperability.

O-RAN RIC is positioned in the middle, where O-RAN RIC is introduced to the existing 3GPP standards. With this addition, it is possible to adopt AI technologies and applications through RIC, while accessing and integrating 3GPP RAN elements. It is also possible to introduce third-party applications through xApps/rApps over RIC. The intention is to adopt AI/ML technologies in a compatible manner with 3GPP standards in 5G and its evolution.

It should be highlighted that the 3GPP standards are adopting AI/ML technologies extensively in a different manner. For example, the concept of intent-based networking, where networks are managed based on intents, a set of expectations including requirements, goals, conditions, and constraints that is typically understandable by humans as well as being interpretable by machines without ambiguity [13]. This concept has been defined in 3GPP Release 17 and further enhancements are being studied in 3GPP Release 18 [13], which is not limited to RAN but also to other parts of the network. Thus, it is not that there is limited AI/ML technologies in the traditional RAN (or non-Open RAN technologies). The key differentiator of O-RAN RIC is the possibility of opening up the AI/ML technology platform of the RAN to the third party (the ones other than the platform provider).

6G is another major milestone for the industry, and 6G vision is already being discussed. Many of the 6G discussions involve AI-native network, and it seems likely to adopt the machine learning and deep learning technologies natively as 5G networks did with cloud technology. Given the consideration that Open RAN paradigm is being included in the 6G vision, the combination of Open RAN and AI-native network is very likely to be the basis of 6G technology. For example, the 6G may include the use of large language model (LLM) to develop and manage, enabling the network to be autonomous.

8.6 Transition to Open RAN

8.6.1 The Path to an Open RAN Compliant Network

The transition from a traditional radio access network to one that aligns with the Open RAN principles is expected to happen at different pace and following a different path for different operators.

At different pace because by the time Open RAN products have reached the market many operators had already started to deploy 5G RAN and therefore they are not likely to replace or upgrade the existing equipment before it is necessary to do so. Ultimately, the pace of transformation will depend on the life cycle of the already deployed base stations and on network expansion.

The statement that operators will follow different paths stems from the consideration that in a few years it will become difficult to draw a clear demarcation between what is considered an Open RAN network and what is not. This is because not only the Open RAN concepts are being transferred and embedded in the mainstream standardization bodies such as 3GPP, but mainly because there is no single criteria that can be used to evaluate the progress of the shift to Open RAN. Is Open RAN defined by the openness of what are now internal interfaces of a base station? Or is it the programmability of the RAN through APIs? Is it rather the number of active suppliers? Or simply the level of virtualization and use of open software?

The term Open RAN is today used by different stakeholders to refer to different types of RAN with different characteristics, and this goes to explain why we see today claims of having reached a certain level of openness in the RAN made by operators even if they have followed very different strategies.

The types of RAN platforms that can be referred to as being Open RAN span from the possibly purest definition of Open RAN which is championed by "new" vendors such as Mavenir and Parallel Wireless to RAN platforms that have some of the features of an Open RAN and are provided by the traditional RAN vendors (Figure 8.10).

Types of RAN	Open interface	Cloud-native	Disaggregated	Multi-vendor	Virtualized
Pure Open RAN	○	○	○	○	○
Pre-integrated Open RAN	○	○		△	○
Appliance-based Open RAN	○	△		△	△
vRAN	△	○		△	○
Cloud RAN		○			○

Figure 8.10 Types of RAN platform that may be considered open.

While greenfield operators, that is new operators who are not burdened by legacy equipment and that are just now starting deploying 5G can start afresh, brownfield operators will be presented with a gamut of alternatives to choose from.

- *Controlled deployment*: Operators may choose to experiment with Open RAN and forging relationships with new vendors in specific parts of their network be it rural areas where traffic is limited, or a campus where they can invite friendly users. Once satisfied with the stability and quality of the RAN, they can then expand it to the full network
- *Outsourcing to system integrators*: Operators can choose to invite a system integrator to manage the deployment of open RAN. While by doing so the operator has less control on vendor selection and openness of the RAN, it can avoid having to develop expertise on new technologies and processes that could be costly and time consuming.
- *Rely on existing vendors to catch up*: Some operators may prefer to maintain the long-standing relationship with their traditional vendors and rely on them to make their products more compliant with the Open RAN philosophy. This strategy may work better for large operators who can use their influence in standards developing organizations as well as purchasing powers to accelerate the adoption of some of the most promising innovations coming from the Open RAN community.

Whichever path is chosen, brownfield operators will have to wrestle with one or more of the many challenges that transitioning to Open RAN will bring such as interworking with existing equipment, selection and management of multiple vendors, life-cycle management, and performance that are discussed more in detail in this section.

Ultimately, for many years to come, the operator RAN is likely to be a mix of all types of virtualized RAN shown in Figure 8.2 that optimizes the TCO and performance.

8.6.2 Coexistence of Deployed RAN with Open RAN Components

Since 4G, it has been especially common for RAN vendors to provide the so-called "Single RAN" solutions whereby a single cabinet hosts all the RAN equipment necessary for 2G, 3G, and 4G access. While single RAN generates efficiency in the fact that passive components are shared, the footprint of the base station is smaller, energy consumption is reduced, and allows vendors to apply proprietary optimizations, it generates a very strong vendor lock-in making adding a new vendor quite challenging. In this situation, operators who desire to transition to Open RAN have often no other choice but deploying the 5G RAN as standalone, alongside the existing vendor. One additional complication is that, as studied in Chapter 4, the typical 5G RAN deployment is in non-standalone mode, meaning that the 5G base station is tightly coupled with the 4G base station and specifically the control plane of the connection utilizes LTE with the user plane utilizing either LTE or NR. Ultimately, most, if not all early deployment of 5G radio has been realized utilizing the same vendor that provided the 4G RAN.

Another important factor that may hinder the transition toward Open RAN is the operation and support system and specifically the migration from the usually proprietary network management system to the more open SMO that O-RAN Alliance is defining. The NMS takes care of FM, configuration, performance management, and security of the base station and is generally optimized to work with a specific base station. Several examples

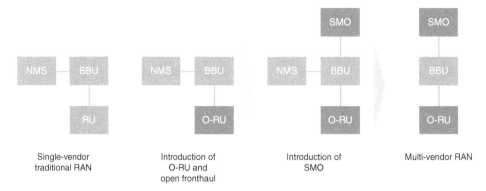

Figure 8.11 Example of migration from proprietary NMS to SMO-based network management.

are provided in literature on how the migration could look like. Figure 8.11 [4] shows how Nokia describes the typical evolution path.

In addition, Vodafone and NTT DOCOMO provide another approach to migrating from traditional RAN to Open RAN with fully O-RAN compliant SMO [14]. While fully O-RAN compliant SMO will harness the full potential of the Open RAN, the interoperability with legacy traditional RAN is a critical bottleneck in the perspective of brownfield operators. In this context, it is rational to retain some of the vendor proprietary interfaces and aspects of the traditional RAN in the transition process. This is what happens in the second phase of Figure 8.12, where the SMO interacts with traditional RAN's element management system (EMS) that to make management of RAN network functions interoperable with SMO and further with O-Cloud. While not all the benefits of Open RAN will be available, such interoperable intermediary architecture can still enable non real-time use cases such as energy efficiency. Of course, this is easier said than done, and the operators will need to take additional considerations such as creating guidelines for operation (e.g. methods to respond to critical but likely situations such as network disconnection) along with providing end-to-end validation environment for RAN solutions and functionalities.

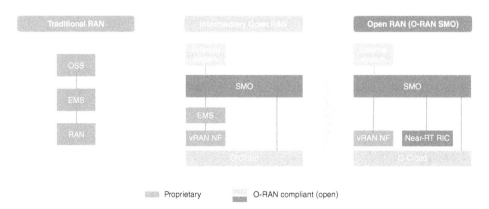

Figure 8.12 Migration from traditional RAN to Open RAN with fully O-RAN compliant SMO. Source: Adapted from [14].

8.6.3 Multi-vendor RAN

Most of the modern mobile radio access networks are already multi-vendor in fact operators typically use radio equipment from two or more different vendors. Each vendor is given responsibility for a certain region of the network and in that region only the vendor equipment is deployed. This setup where the overall RAN is in fact the combination of different RANs, each operating in a specific geographic area implies that interaction between base stations of two different vendors, can only happen at the border between the regions. In reality, even if from 4G onward 3GPP has specified an interface between base stations (X2) and techniques exist to improve the throughput of a data connection by using more than one base station (such as CoMP), these features are not used if base stations do not belong to the same vendor. The implication is that operators are quite used to dealing with multiple vendors, whereas vendors are not used to work with each other. The type of multi-vendor RAN envisaged by Open RAN proponents, however, is quite different in the fact that different vendors are expected to contribute different parts of the base station and any base station is expected to interoperate with any base station. In this type of multi-vendor environment, vendors are expected to establish strong cooperation and collaborate to troubleshoot possible compatibility issues between their components.

8.6.4 Open RAN Equipment Lifecycle

In line with the Open RAN concept, base stations are set to transform from monolithic pieces of equipment provided and managed by a single vendor to systems that are assembled using components from several different vendors and connected using open interfaces. While the benefits of such "mix-and-match" approach are clear, the operator deploying such base stations will need to evolve its approach to the management of the lifecycle.

Currently, operators are used to deploy single-vendor base stations that have been fully tested and certified by the supplier in a turn-key solution fashion. When the base station is, however, assembled using components from different vendors, there will be a burden on the operator, or a third-party integrator that the operator may appoint, to verify that all the modules are correctly interoperating and can be successfully managed by the SMO. As the number of vendors of Open RAN vendors increase, so do the possible combinations and it may very well happen that the setup of the base station the operator has decided upon has never been deployed by other operators. While for monolithic base stations, the troubleshooting is a responsibility of the vendor which can draw from experience with other customers and is in control of the full system, if a problem occurs with an Open RAN compliant base station it is the burden of the operator to perform a root cause analysis to determine which component (or components) is causing the fault and ask the corresponding vendor to fix it. Troubleshooting is also likely to require close cooperation between the vendors that provide the components.

Besides the initial configuration at the time of deployment, there may also be challenges in the base station feature roadmap. Traditional vendors can offer a clear evolutionary path for their products as well as a catalog of features that can be added to their product ensuring minimum downtime and well-defined timeline. With Open RAN base stations, introducing new features may require that the vendors synchronize their roadmap and that the upgrades are tested in advance in the operator labs (provided the operator has one).

8.6.5 Open RAN Performance

Initially, it is unlikely that the performance of an Open RAN base station will match the performance of a single-vendor base station. This is because traditional vendors have the control of the full base station system and have spent many years developing strategies and optimizations on top of the standards or in areas where the functionality is not strictly specified and left to implementation. For example, the packet scheduler, which is responsible for determining the order in which data packets are transmitted to the users served by the base station, is an area where vendors can differentiate as its inner working is not specified by 3GPP. Traditional vendors can also use the internal interfaces in the base station to exchange proprietary messages.

Another area where traditional vendors still have the edge is in the use of hardware accelerators either in the form of multipurpose field programmable gate arrays (FPGA) or specialized application-specific integrated circuits (ASIC) that can greatly boost the performance of the base station and increase the energy efficiency compared to software-based solutions that are preferred by Open RAN advocates.

Over time, the performance gap is likely to be bridged by new vendors and Open RAN base stations are expected to outperform traditional ones, but when considering the transition to Open RAN, this is an important consideration.

8.7 Moving Toward the Future Access Agnostic Network: Nonterrestrial Open RAN Scenarios

8.7.1 General

As the 3GPP specification work progresses for 5G, new items and technologies are taken under consideration, and some of them might have a significant impact on the earlier 3GPP functionalities and performance. One of these emerging technologies is the nonterrestrial network (NTN) concept. It is defined for the first time in the 3GPP Release 17, and the consecutive releases present further definitions to evolve the concept. As the radio interface of NTN will include considerably longer distances, it impacts, e.g. to the HARQ procedure; instead of 16 frames, the 3GPP has reasoned that NTN benefits from 32 frame span for the HARQ acknowledgements to ensure the acknowledgment can be delivered within the expected time window. NTN therefore merits consideration in the context of Open RAN, and this section outlines the considerations arising from NTN and the impacts on Open RAN and its respective measurements.

8.7.2 Frontend

NTN can benefit from revised frontend definitions as the CPRI in 4G, or eCPRI in 5G, might not suit optimally to the space communications. The digital intermediate frequency interoperability (DIFI) consortium is an independent group of space companies, organizations, and government agencies from around the world that promote an open and simple standard to facilitate interoperability of ground systems supporting space operations.

Digitizing analog intermediate frequency (IF) with the DIFI open standard allows new ground station architectures based on cloud services. DIFI's mission is to accelerate the digital transformation of the satellite industry.

Interpreting the advances from [15–18], currently, the DIFI v1.0 standard is still in rather early stage. It specifies some VITA 49.2 packets over user datagram protocol (UDP) and 802.3 standards. In the future versions, it is expected that DIFI will specify more concrete transport features for reliable packet delivery. Specifically, the session layer will need to address reliable transport of UDP packets.

8.7.3 Polarization

As stated in [19], the polarization of an electromagnetic wave is defined by the orientation of its electric field vector. The signal polarization can be either linear or circular in satellite systems. Typically, the Ku-band relies on linear polarization (LP). In practice, the implementation of mobility solutions is challenging on Ku-band, though. The most satellite signals using Ka-band rely on circular polarization (CP), that is, the electric field rotates clockwise or anticlockwise looking along the direction of propagation. Once the receiving antenna is aligned for maximum signal by pointing along boresight to the transmitting satellite, CP avoids the need for rotation of the feed to any transverse angle [20].

According to [21], the 3GPP TS 38.821 states that another issue raised several sources is on polarization mode configuration and signaling. In NTN networks, neighboring cells may use different polarization modes to mitigate intercell interference, and there may be UEs with different antenna types. Some UEs may be equipped with linearly polarized antennas, while some other UEs may be equipped with circularly polarized antennas. Three sources proposed that it is beneficial to signal the polarization mode for NTN in certain scenarios.

As per [22], circularly polarized antennas, apart from being used in wireless communication systems, are useful in satellite communications as they reduce multipath effects besides mitigating polarization mismatch between the transmitting and receiving antenna thereby reducing polarization loss. CP antennas are advantageous as they can increase the efficiency of the radio communication system due to their ability to reduce the undesired fading effects caused due to multipath effects, realizing duplex channels for frequency reuse, and to enhance the capacity of the overall system as well as the quality of the link.

As per [22], in nonline of sight (NLOS) scenarios, such as mine tunnels, cavities, and stopes, it is not necessary to match the polarization. Furthermore, as per Ref. [23], Ericsson and Huawei have proposed that NR NTN should support configuration of DL and UL transmit polarization including right-hand and left-hand CPs (RHCP, LHCP). Interested readers are encouraged to refer to further information on polarization in Ref. [24].

8.7.4 MIMO

As per [25], related to MIMO for Satellite Communication Systems, MIMO using multiple satellites is an attractive application at lower frequency bands, especially at UHF SATCOM to enhance capacity by orbital frequency reuse. A multiuser (MU) MIMO spatial multiplexing technique is proposed in publication to enhance the channel capacity of a HiCapS system in a constrained geographical region by enabling frequency reuse.

The main advantage of the MU-MIMO technique is that each user requires only a single antenna terminal and it is applicable in LOS channels. The channel capacity can be linearly increased utilizing multiple spot beam antennas on a single satellite serving the considered geographical region. In 3GPP studies, MU-MIMO has been suggested as a potential solution to achieve frequency reuse in SATCOM with a scattered user distribution. Reference [25] expands on the MU-MIMO and spatial multiplexing techniques to show how this can be applied to SATCOM to increase the capacity of a HiCapS system.

8.7.5 Open RAN in Satellite Communications

Figure 8.13 depicts the 3GPP-defined architecture for NTN satellite link as per the technical report TR 38.821 [26]. This scenario refers to a typical nonterrestrial network based on transparent payload in such a way that the gNB is located to the ground.

Figure 8.14 depicts the same for regenerative payload. Regenerative satellite can be with or without inter-satellite link (ISL). In both cases of regenerative setup, the satellite houses gNB for payload processing.

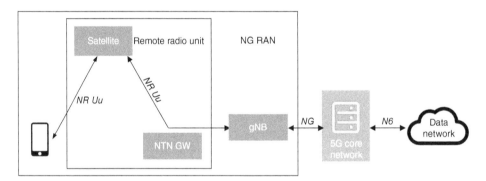

Figure 8.13 Networking-RAN architecture with transparent satellite.

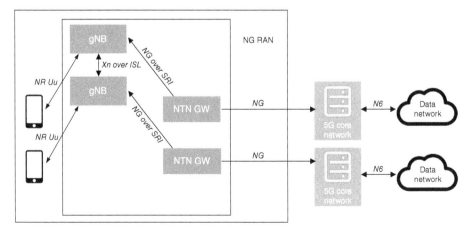

Figure 8.14 Regenerative satellite with ISL, gNB processed payload.

8.7.5.1 NG-RAN Impacts, Transparent Satellite

As the 3GPP TR 38.821 states, there is no need to modify the NG-RAN architecture to support transparent satellite access [26]. Nevertheless, compared to the terrestrial networks, the considerably longer propagation delay of the space radio interface, that is, uplink and downlink between UE, satellite, and gateway needs to be considered. As an example, for scenarios involving geostationary orbit (GEO) satellites, the NR-Uu timers need to be extended so that they cope with such a long delay of the feeder and service link. Although the low earth orbit (LEO) scenarios are less of a concern for the extended delay, the LEO scenario involving ISL, the delay would be a result of the satellite radio interface (SRI) (feeder link) and at least a single ISL.

8.7.5.2 NG-RAN Impacts, gNB Processed Payload

As reasoned in the 3GPP TR 38.821, the NG Application Protocol timers may need to be extended to compensate the long delay occurring for the feeder link communications [26]. There can occur up to several hundreds of ms delays on NG for scenarios involving GEO satellites compared to terrestrial networks. This latency can be compensated for the UP and CP by designing adequate implementation.

As is the case with the transparent satellite communications, the overall delay for the ISL scenario of LEO satellites needs to consider the SRI (feeder link) and at least one ISL.

8.7.5.3 Impacts of the Open RAN

The Open RAN concept can be used as a part of the satellite communications so that it forms a logical component within the 3GPP NTN architecture. This scenario would look as depicted in Figure 8.15. This scenario refers to "gNB-DU processed payload" as defined in the 3GPP TR 38.821 [26]. The NG-RAN logical architecture with CU and DU split, as per 3GPP TS 38.401, forms the baseline for the NTN scenarios.

In this scenario, the satellite regenerates the signals it receives from Earth so that the NR-Uu radio interface functions as the service link between the satellite and the UE, whereas the SRI is the feeder link between the satellite and ground NTN gateway for transporting the F1 protocol. The SRI is thus a 3GPP-specified transport link, and the NTN GW works as a transport network layer node supporting the needed transport protocols. The satellite may also provide intersatellite links in the regenerative model.

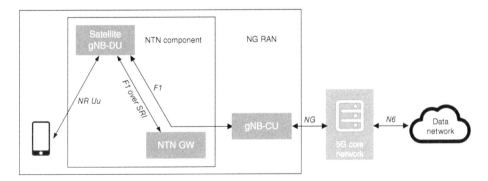

Figure 8.15 NG-RAN with a regenerative satellite based on gNB-DU.

The split model in this scenario has the DU on board of each satellite so that different satellites may be connected to the same CU on ground. If the satellite hosts more than one DU, the same SRI will transport all the corresponding F1 interface instances.

As stated in 3GPP TR 38.821, also in this scenario, there will be some NG-RAN impacts. As an example, the radio resource control (RRC) and other Layer 3 processing are terminated in the gNB-CU on ground and are subject to stricter timing constraints [26].

The use of this architecture option for LEO or GEO systems may impact current F1 implementation, requiring for example, extended timer values. As is the case in previous scenarios, the impact for GEO satellite communications is greater than for LEO satellites.

8.7.5.4 Impacts on Measurements

Being agnostic to the actual air interface connectivity, the Open RAN concept per se does not change the basic principles of the field measurements. The same performance indicators are still valid, thus, as well as the measurement methods. Examples of these are the received power level, bit error rate, and quality indicators. Furthermore, the KPIs can be constructed from the measurement attributes, which allows means for understanding the radio performance of different scenarios. Such KPIs can include the more detailed analysis of, e.g. the received power level ranges per the respective modulation schemes.

To ensure the correct setup of the NTN, the involved stakeholders need to verify the correct functioning and expected performance figures under the practical scenarios. The verification can be done through field tests. The key performance indicators reveal possible issues in network parameter values and help verify that the network's functional parameters are correctly updated. An example of this is the verification of the correct functioning of extended timers and radio elements that consider correctly the extended HARC acknowledgements.

References

1 5G Americas, "The Evolution of Open RAN," 2023.

2 S. Munir, "Introduction to Open RAN," 2023.

3 GCC Open RAN Consortium, "Open RAN for Brownfield Operators Challenges and Opportunities," 2022.

4 Spirent, "How to test Open RAN," Spirent, 2022.

5 Spirent, "Will Your Open RAN Deployment Meet User Expectations?," Spirent, 29 March 2023. [Online]. Available: https://www.lightreading.com/webinar.asp?webinar_id=2217. [Accessed 7 September 2023].

6 Vodafone, KeySight, "Open RAN Handbook," Vodafone, KeySight, 2023.

7 DOCOMO, "5G Open RAN Ecosystem Whitepaper," DoCoMo, June 2021. [Online]. Available: https://www.docomo.ne.jp/english/binary/pdf/corporate/technology/whitepaper_5g_open_ran/OREC_WP_20230517095959.pdf. [Accessed 4 October 2023].

8 O-RAN Alliance, "Overview of Open Testing and Integration Centre (OTIC) and O-RAN Certification and Badging Program – White Paper, April 2023," O-RAN Alliance, April 2023. [Online]. Available: https://www.o-ran.org/resources. [Accessed 4 October 2023].

9 O-RAN Alliance, "Testing & Integration," O-RAN Alliance, [Online]. Available: https://www.o-ran.org/testing-integration#learn-otic. [Accessed 3 October 2023].

10 O-RAN Alliance, "Overview of Open Testing and Integration Centre (OTIC) and O-RAN Certification and Badging Program," O-RAN Alliance, 2023.

11 J. T. Penttinen, E. Tsakalaki and P. Amin, "Optimization of LTE-A," in *LTE-A Deployment Handbook*, John Wiley & Sons, Ltd, 2016, pp. 293-337.

12 J. Choi, "AI-Native Open RAN for 6G," 2023.

13 R. Xu, M. Scott and S. Mwanje, "Intent Driven Management," 27 07 2023. [Online]. Available: https://www.3gpp.org/technologies/intent. [Accessed 29 09 2023].

14 Vodafone, NTT DOCOMO, "SMO Deployment Challenges in an Open RAN Environment," 2023.

15 Digital Intermediate Frequency Interoperability (DIFI) Consortium, "Digital Intermediate Frequency Interoperability (DIFI) Consortium," 1 October 2023. [Online]. Available: https://dificonsortium.org/#:~:text=The%20mission%20of%20the%20Digital,IF%20signals%20and%20helps%20prevent.

16 MIcrowave Journal, "The Role of Satellites in 5G Networks," 13 May 2020. [Online]. Available: https://www.microwavejournal.com/articles/print/33942.

17 P. P. Juan Deaton, "Digital Transformation of Satellite Communication Networks," March 2022. [Online]. Available: https://csiac.org/wp-content/uploads/2022/03/CSIAC_SOAR_Digital-Transformation_Deaton-004.pdf.

18 C. R. Andres, "AWS joins the Digital IF Interoperability (DIFI) Consortium," 26 May 2022. [Online]. Available: https://aws.amazon.com/blogs/publicsector/aws-joins-the-digital-if-interoperability-difi-consortium/.

19 J. Ness, "Technically Speaking: Switching Between Satellites with Different Polarization Types," September 2017. [Online]. Available: http://www.milsatmagazine.com/story.php?number=429273747#:~:text=Most%20satellite%20signals%20at%20Ka,left%20hand%20(LHCP)%20respectively.

20 Avnet Silica, "The Open RAN System Architecture and mMIMO," 21 November 2021. [Online]. Available: https://www.avnet.com/wps/portal/silica/resources/article/open-ran-system-architecture-and-mmimo/.

21 Ericsson, "Using 3GPP technology for satellite communication," 1 June 2023. [Online]. Available: https://www.ericsson.com/en/reports-and-papers/ericsson-technology-review/articles/3gpp-satellite-communication.

22 V. C. M. B. e. a. U. Ingle, "Circularly Polarised Reconfigurable Antenna in 5G Application: A," 5 May 2021. [Online]. Available: https://digitalcommons.unl.edu/cgi/viewcontent.cgi?article=10638&context=libphilprac.

23 M. Inc, "3GPP R1-200XXXX-AI_8_4_4-Summary of 8.4.4 Other Aspects of NR-NTN," 3GPP, Sofia Antipolis, France.

24 JEM Engineering, "Microstrip Antennas: The Basics," October 2020. [Online]. Available: https://jemengineering.com/blog-intro-to-antenna-polarization/.

25 TechLte World, "5G-NTN (Non-Terrestrial Networks) Overview," 17 September 2023. [Online]. Available: https://www.linkedin.com/pulse/5g-ntn-non-terrestrial-networks-overview-techlte-world/.

26 3GPP, "TR 38.821 V16.2.0: Solutions for NR to support non-terrestrial networks (NTN), Release 16," 3GPP, March 2023.

Index

Note: *Italicized* and **bold** page numbers refer to tables and figures, respectively.

Open RAN Explained: The New Era of Radio Networks, First Edition.
Jyrki T. J. Penttinen, Michele Zarri, and Dongwook Kim.
© 2024 John Wiley & Sons Ltd. Published 2024 by John Wiley & Sons Ltd.